Architecture and Fire

Architecture and Fire

A Psychoanalytic Approach to Conservation

Stamatis Zografos

UCLPRESS

First published in 2019 by
UCL Press
University College London
Gower Street
London WC1E 6BT

Available to download free: www.uclpress.co.uk

A CIP catalogue record for this book is available from The British Library.

ISBN: 978-1-78735-372-5 (Hbk.)
ISBN: 978-1-78735-371-8 (Pbk.)
ISBN: 978-1-78735-370-1 (PDF)
ISBN: 978-1-78735-373-2 (epub)
ISBN: 978-1-78735-374-9 (mobi)
ISBN: 978-1-78735-375-6 (html)
DOI: https://doi.org/10.14324/111.9781787353701

Contents

List of figures

Preface

A number of scholarly voices in recent years have expressed interest in exploring the collaboration between two disciplines that are not traditionally studied alongside one other: architecture and psychoanalysis. This book celebrates this emerging discourse by offering a reading of architectural conservation through Freudian psychoanalysis, according to which key theoretical paradoxes and inconsistencies associated with the former can be reconsidered. This approach benefits from the creative and critical potential that emerges 'between *and* across'[1] architecture and psychoanalysis when the two are brought together and examined in close proximity. As the architectural theorist Jane Rendell suggests, the essence of interdisciplinarity is to challenge the 'edges and borders'[2] of the disciplines in question, so in this sense, how is architectural conservation related to psychoanalysis? The psychiatrist Cosimo Schinaia rightly points out that 'we live inside architectural structures, for instance our homes, but at the same time they live inside our minds: in dreams, for example, we can build architectural structures, modify them, or destroy them'.[3] In this light, both disciplines deal with space, which is defined by boundaries that can be alternately simple and difficult to identify.

Architectural conservation deals with clear and obvious boundaries. It preserves the boundaries that architecture builds. Walls and facades are architectural boundaries that distinguish the habitable from the non-habitable, the internal from the external, the intimate from the distant and the private from the public. Architectural boundaries set limits. They make distinctions but at the same time enable the simultaneous existence of discreet entities, of our private selves, within a greater whole, the city itself. By default, walls and facades – architectural boundaries – are multi-layered in the sense that they carry and convey layers of meaning. They also define the levels of transparency that control how the inside communicates with the outside. Conservation seeks to preserve these boundaries and to protect our space and privacy against the unfolding of time. Thus architecture and conservation deal closely with space and time.

The boundaries that define the space of psychoanalysis are not particularly clear at first sight. The discipline of psychoanalysis scrutinises the unconscious, which has no tactile existence. It occupies no place and has no temporality. Unlike architecture and conservation, the unconscious carries no spatial and temporal characteristics. So how can we talk about space or boundaries in psychoanalysis? Boundaries in this discipline commonly emerge during its practice. Psychoanalysis deals with the inside, the psychic world, and the outside, the physical world. Repetitive therapy sessions gradually build an intimate, private space between the individual and the other, the analysand and the analyst, the individual and the world. This intimate space is occupied with memories, dreams, obsessions, anxieties and compulsions. In this sense, psychoanalysts, architects and conservationists are all creators of intimate spaces.

This book is not just about architectural conservation. There is a parallel thematic thread running throughout this book that deals with fire. In *The Psychoanalysis of Fire*, the philosopher Gaston Bachelard admits that our knowledge of the element today is not only limited but also taken for granted. Knowledge of fire comes to us in the form of a prohibition, from our elders, while scientific books rarely make reference to it.[4] The present book is an attempt to compile information about fire – both as an element and a concept – through the engagement with sources from diverse disciplines, aiming to illuminate our scattered and obscure knowledge of it.

This research on fire focuses on a single aspect, which, as the title of the book suggests, is its interaction with architecture. The topic of architecture and fire has sadly received huge attention in recent years, particularly in the aftermath of tragic events that have scarred modern cities. The Grenfell Tower fire catastrophe that took place in London in June 2017, the fire that devastated for the second time Mackintosh's masterpiece building at Glasgow School of Art in June 2018, the obliterating wildfires in the outskirts of Athens in July 2018 and the devastating fire at the National Museum of Brazil in Rio de Janeiro in September 2018 are four very different fire incidents that have not only shocked audiences around the globe but also awakened an interest in reassessing our overall approach to the element. These tragic incidents confirm that the risk of fire catastrophe is always imminent, regardless of the 'preparedness' of contemporary societies to fight them. Yet they also highlight the necessity for the urgent implementation of additional measures that can reduce, if not eliminate, this apparent risk. This book therefore contributes to this newly emerging discourse, as it investigates how architecture, the urban landscape and societies together remember and respond to the continual risk of fire.

Notes

1. Jane Rendell, 'Working Between and Across: Some Psychic Dimensions of Architecture's Inter– and Transdisciplinarity', *Architecture and Culture*, 1: 1 (2013), 136.
2. Rendell, 'Working Between and Across', 131.
3. Cosimo Schinaia, *Psychoanalysis and Architecture: The Inside and the Outside*, translated by Giuseppe Lo Dico (London: Karnac, 2016), xxiii–xxiv.
4. Gaston Bachelard, *The Psychoanalysis of Fire*, with a preface by Northrop Frye (Boston: Beacon Press, 1964), 2–3.

Acknowledgements

This book is the outcome of a long and convoluted process that was inspired by my Master's dissertation in Cultural Studies on the topic of 'fire and forgetting'. This topic gradually turned into a PhD thesis that was submitted in 2012 to the Centre for Cultural Studies at Goldsmiths University of London. During this process and until this publication, I have hugely benefited from the comments, advice and assistance of many colleagues and friends. I am therefore very grateful to the following individuals: Edwina Attlee, Eva Bachmann, Iain Borden, James Burton, Ben Campkin, Paola Crespi, Nicola Collett, Marie Dumas, Saskia Fisher, Uli Gamper, Carla Garcia, Sebastian Hicks, Lorens Holm, John Hutnyk, Astero Klampatsa, Scott Lash, Nela Milic, Carlos Molina, Lee Palmer, Elena Papadaki, Stellios Plainiotis, Liana Psarologaki, Stefania Scarsini, Bernard Stiegler, Sotirios Varsamis and Gary Wragg. Huge thanks also to Chris Penfold, the Commissioning Editor at UCL Press, for commissioning this book and offering his invaluable advice throughout the publishing process.

I would also like to thank enormously the following people and organisations for their kind contributions with images for this book: Alfredo Jaar, Caruso St John Architects, Hélène Binet, Jack Wates, Lois Weinberger, Louis Benassi, Mae Architects, Philippe Rahm Architectes, Riccardo Vincentini, Rocio Ayllon, Slider Studio, Suzanne Ludwig and Querkraft Arkitekten.

I would also like to acknowledge the UCL Bartlett School of Architecture's Architecture Research Fund. Without their financial support, certain aspects of this book would not have been possible.

The research of this book was partly conducted at the library of the (now relocated) London Fire Brigade Museum in Southwark and the London Metropolitan Archives. I would like to acknowledge the generous assistance of the staff during my research.

Finally, I would like to thank my family, and above all my parents, for their boundless support during this process.

All things change to fire,
and fire exhausted
falls back into things.
Heraclitus, Fragment XXII

1
Introduction

> We can no longer afford to take that which was good in the past and simply call it our heritage, to discard the bad and simply think of it as a dead load, which by itself time will bury in oblivion. The subterranean stream of Western history has finally come to the surface and usurped the dignity of our tradition.
>
> Hannah Arendt, *The Origins of Totalitarianism* (1951)

> We speak so much of memory because there is so little of it left.
>
> Pierre Nora, *Between Memory and History: Les Lieux de Mémoire* (1989)

On 12 May 2018, the East London film centre Close-Up organised a screening of film triptychs by Louis Benassi called *After '68: There Are Three Ways to Resist but the Enemy Is One*. One of the triptychs was *Black Umbrella* (2011), a 16mm film installation depicting the burning of the Crystal Palace in 1934, the flying-bomb raids in central and East London in the 1940s, and the fire at the Houses of Parliament in 1958. All three films are made with discarded archive material that was discovered accidentally in a disused fire station in London. The triptych was screened in a horizontal arrangement; (fig. 1.1) a fourth film was superimposed on it, showing a woman walking through a city (fig. 1.2) holding a black umbrella.[1]

Benassi's superimposition of this iconic image of a person carrying a black umbrella is inspired by a common modernist experimental iconography, such as Magritte's surrealist paintings of people with hats and umbrellas or Francis Bacon's paintings produced between 1945 and 1946. Benassi comments on his work:

> The umbrella acts as a portable architectural dome or roof providing shelter from the rain or sun, however in the context of the triptych the object is shielding 'our' young woman from the explosive, life threatening splinters produced by the flying bomb, the flying bomb, which incidentally could be seen as a metaphor for the heartless architects of displacement.[2]

Figure 1.1 Stills from Louis Benassi's film triptych *Black Umbrella*. © Louis Benassi.

Figure 1.2 Still from the superimposed film in *Black Umbrella*. © Louis Benassi.

Black Umbrella touches on themes central to this book, including the role of archives in the preservation of memory and the destruction of buildings by fire. It also signals the breadth of contemporary discourse on the concept of the archive, which is one of the principal contexts for this book.

This is a book about architectural conservation. Today conservation receives unprecedented attention as a direct reflection of a more general interest in memory that has been apparent during the last few decades. During this period, we have witnessed an increasing desire for the perpetuation of memories from the past, a desire that applies to every aspect of our culture. There is widespread investment in the construction of new museums and the restoration or extension of existing ones, in the erection of new memorials, in building new libraries and conducting genealogical and biographical research, in the organisation of commemorative events and the revival of nearly extinct traditions, in the exploration of dark and difficult periods of repressed history that are now being brought to the surface, and so on. This desire to perpetuate memories does not manifest itself only on a collective level but also on an individual one. Regardless of age, origin or social group, it is a common practice for people to record and preserve personal moments on a daily basis. This is possible through the use of digital and analogue recording technologies, such as cameras, DVDs, CDs and films, as well as through the Internet and social media. In general, we are experiencing a period during which remembering has obtained an immense significance whereas forgetting is rarely considered an option.

Reflecting on this obsession with memory, Frances A. Yates, a notable scholar on memory issues, has expressed that 'we moderns have no memory at all'.[3] One needs only to consider that nowadays, remembering is a responsibility assigned almost exclusively to computers, which, as modern mnemonic tools, record, save and retrieve data and thus replace human memory. Commenting on this apparent replacement, cognitive scientists claim that computer memory is virtually the same as human, the only difference being that the former does not fail. Before their invention, the work of computers was done by humans. Therefore, as the media theorist Warren Sack explains, 'computer memory seems to be a good model of human memory because computer memory was modelled on human memory!'[4] Nevertheless, compared to the human process of remembering, the function of computers comprises only part of the human process as 'human memory has become self–externalised: projected outside the rememberer himself or herself and into non–human machines.'[5] The general attitude today favours remembering over the possibility of forgetting.

The French historian Pierre Nora attributes this contemporary tendency towards the perpetuation of memories to the *acceleration of history*, which has brought about the complete collapse of real memory and its subsequent equation with history. In his extensive work on national French memory called *Realms of Memory*, Nora explains how real memory has gone through many stages of degradation.[6] Most crucial is the eradication of peasant culture, which once operated as a repository of collective memory. Peasant culture transmitted this collective memory, which for Nora is real memory, through the ritualistic, repetitive practice of quotidian activities.[7] Contrasting memory and history, he writes:

> Memory is life, borne by living societies founded in its name. It remains in permanent evolution, open to the dialectic of remembering and forgetting ... History, on the other hand, is the reconstruction, always problematic and incomplete, of what is no longer. Memory is a perpetually actual phenomenon, a bond tying us to the eternal present; history is a representation to the eternal past ... Memory is absolute, while history can only conceive the relative.[8]

As a consequence of the acceleration of history, Nora remarks that the present is no longer the link between the past and the future. Traditionally memories of the past were critically selected and organised in anticipation of the future, which gave a meaningful sense of duty in the present. Nowadays, however, the lack of vision for the future urges a frantic investment in the preservation of every memory possible from the past, which Nora calls *the duty to remember* (*devoir de mémoire*).[9] Nora adds that this upsurge in memory is also caused by the democratisation of history, as minority groups, following decolonisation, can now reaffirm their identity by uncovering and establishing their own past.[10]

A different view suggests that this tendency to record every single memory from the past reflects a prominent feature of globalisation, namely change, which comes to question, and often erode, traditionally established socioeconomic and cultural structures. According to the British sociologist Anthony Giddens, globalisation is a dialectical process and is 'defined as the intensification of worldwide social relations which link distant localities in such a way that local happenings are shaped by events occurring many miles away and vice versa'.[11] For the German sociologist Ulrich Beck, globalisation is associated with the period of post-modernisation, during which 'we are witnessing not the end but the beginning of modernity – that is, of a modernity beyond its classical industrial design',[12] and this is what he calls *reflexive modernisation*. An unavoidable consequence of this late modernity is risk, which in previous eras was essentially personal, but now manifests itself on a global scale. For instance, in the current time we are imminently threatened by ecological risks, such as environmental pollution, which affect every form of life on the planet. This is because risks have a boomerang effect, which assures that even the source that generates them will be equally harmed. Risks cannot be restricted either within nations or among the poor, for as Beck points out, 'poverty is hierarchic, smog is democratic'.[13] In this sense, risks are both local and global; thus the risk society we live in is a *world risk society*. Due to their scale and magnitude in the era of globalisation, risks are not easily controllable and calculable. The risk society, for both Giddens and Beck, has a salient characteristic, which is its preparedness to deal with problems and hazards occurring as a result of reflexive modernisation. As Beck claims, 'risks have something to do with anticipation, with destruction that has not yet happened but is threatening … risks signify a future which is to be prevented'.[14] Thus in current risk societies, the past no longer determines the present but it is the future, or rather, the potential threats to the future, that determine the present actions of the risk society. This anticipation of an uncertain future and the fear of potential irreversible destruction bring about an increasing, sometimes even frenzied reaction, which, although it can probably not help alleviate problems associated with globalisation,[15] explains why there is an increasing investment in the perpetuation of memories from the past and present.[16]

The heritage studies scholar Rodney Harrison equally identifies an exponential increase in the investment in memory, which manifests itself with the stockpiling of material traces from the past to the extent that 'almost everything can be perceived to be "heritage."'[17] He calls this phenomenon the *heritagisation of society* and warns that this overwhelming tendency to accumulate memory can bring about the reverse, unwelcome result, where all heritage becomes worthless.[18]

The *duty to remember* manifests itself so frequently and so intensely that nearly every aspect of the world must be archived. The potential creation of an all-encompassing archive is reminiscent of Jorge Luis Borges's imaginary concept of 'The Aleph'. In this short story, Borges approaches a corner of his friend's cellar and, looking up in darkness, he observes an Aleph. This Aleph, the diameter of which is no wider than an inch, is 'one of the points in space that contains all points … [It is] the place where, without admixture or confusion, all the places of the world, seen from every angle coexist'.[19] Staring at the Aleph for a few moments, Borges witnesses 'the inconceivable universe',[20] which he finds impossible to describe. This universe appeared before his eyes in a simultaneous occurrence and so a description of it can never be complete, for language develops in a successive order.[21] Having experienced the Aleph, he realises that life is pointless because everything looks familiar. But later, he confirms that 'fortunately, after a few unsleeping nights, forgetfulness began to work in me again'.[22] Borges's approach to memory favours a balanced analogy between remembering and forgetting, which in essence contradicts the archival obsession of our times.

A similar approach to memory is shared by the German philosopher Friedrich Nietzsche. In his essay *On the Uses and Disadvantages of History for Life*, published in 1874, Nietzsche writes:

> Consider the cattle, grazing as they pass you by: they do not know what is meant by yesterday or today, they leap about, eat, rest, digest, leap about again, and so from day to day, [are] fettered to the moment and its pleasure or displeasure .... A human being may well ask [such an animal]: 'Why do you not speak to me of your happiness but only stand and gaze at me?' The animal would like to answer, and say: 'The reason is that I always forget what I was going to say' – but then he forgot this answer too, and stayed silent: so that the human being was left wondering. But he also wonders about himself, that he is not able to learn to forget and that he always hangs onto past things. No matter how far or how fast he runs, this chain runs with him. It is something amazing: the moment, in one sudden motion there, in one sudden motion gone, before nothing, afterwards nothing, nevertheless comes back again as a ghost and disturbs the tranquillity of each later moment. A leaf is continuously released from the roll of time, falls out, flutters away – and suddenly flutters back again into the man's lap. For the man says, 'I remember', and envies the beast, which immediately forgets and sees each moment really perish, sink back in cloud and night, and vanish forever.[23]

In this essay, Nietzsche suggests that a cow lives in happiness because it does not remember. And precisely because of this forgetfulness, there is no chance it can confirm its happiness, as it cannot recall its previous state. Therefore on the one hand happiness is achieved through an absence of memory, and on the other it is taken away in the same instance. Through this example Nietzsche's intention is to emphasise the liberating power of what he calls 'active forgetfulness',[24] which is a possibility only for humans and not for other animals. The grand theme of his book is the abandonment of certain aspects of the past, as the latter returns as a ghost and disturbs the peace of a later moment. His suggestion is a critical discourse on the past, which intends to keep only the knowledge and experiences that are beneficial for current and future life whereas all unnecessary and disadvantageous elements should be left behind. Thus, active forgetting is selective remembering. Both Borges and Nietzsche share a closely related approach to memory. Both of them perceive memory's counter-manifestation of forgetting to be liberating, and Nietzsche makes an additional point, which highlights a potential glitch inherent in our contemporary archiving culture. This is the fact that the *duty to remember* must be selective, or at least not totally inclusive, for '[w]hen an archive has to collect everything ... it will succumb to entropy and chaos … . In such cases the system begins to swing back and forth so violently that it finally collapses.'[25]

Although this book challenges our obsession with the preservation of past memories quite broadly, its focus is more specific. It deals with architectural conservation. Conservation research is a common presence in most Western nations' cultural agendas, receiving considerable governmental attention and funding, ongoing professional engagement and widespread public support. In England, Historic England is 'the public body that helps people care for, enjoy and celebrate England's spectacular historic environment'.[26] In 2016 they published their *Three Year Corporate Plan 2016–19* in which they outline their evolving mission, their vision and ambition for conservation research, the resources they have available and how their progress will be assessed.[27] Other conservation research initiatives are supported by the Framework Programmes that have been funded by the European Union since the last decade of the twentieth century. The most famous promote pan-European collaborations, which in the past focused on architecture, but today their range is more inclusive and diverse.[28]

When the historical significance or value of a building is established, it is commonly designated as listed, which means that specific qualities and aspects of its past must be conserved and/or restored. On this occasion, conservation practice aims to prolong a building's current condition while restoration practice reinstates earlier stages of the building's life. In the last two decades, the scope and significance of architectural conservation have expanded broadly for various reasons. Previously it was only monuments from the eighteenth century that were preserved, but today it is nearly every built structure from that century.[29] Architectural conservation is no longer restricted to the preservation of individual buildings but can also include entire neighbourhoods, villages or even whole cities, such as the case of Venice and its lagoon.[30] In the United Kingdom, the establishment of conservation areas is a relatively frequent phenomenon. The first conservation area was designated in 1967 and currently there are more than 8000 designations.[31]

Another recent addition to the conservation agenda is the protection of cultural landscapes,[32] like the Royal Botanic Gardens in Kew, or the protection of historical fakes like Dresden's city centre, restored after its major destruction during World War II.[33] Conservation also deals with structures that are not only disused but have also been left to deteriorate to such an extent that they are beyond recognition. The artist, architect and preservationist Jorge Otero-Pailos, commenting on the conservation of one such structure, a waterfront pier in Manhattan, writes that:

> when an object fails to satisfy the traditional categories of historical significance yet there is still public pressure to preserve it, preservationists are forced to confront that which they habitually repress: meaninglessness .... Preservation provides the illusion that those buildings, which are condemned to the horror of disappearing, are beautiful in their act of disappearance.[34]

Defending the broadly inclusive range of contemporary conservation, the American conservationist James Marston Fitch stresses the necessity of preserving the prototype for the benefit of future generations.[35] Maintaining authenticity is imperative because, as the conservationist John H. Stubbs claims, 'there is no substitute for direct observation of the real thing, which represents the shortest distance in time and space that an object [or a place] and a viewer can have'.[36] The integrity of prototypes is commonly threatened by various processes that take place in urban environments. Apart from the obvious impact of time and nature on buildings, architecture depends highly on the human factor too. Stubbs identifies the following three categories as threats to architecture by humans: 'ancillary effects of modern life (pollution, economic, religious, social or life-style changes); wilful calculations (vandalism, war or terrorism-related destruction); and oversights (ignorance, neglect, profligate use of natural resources or insensitive or inadequate work)'.[37] Of all these threats, the consequences of modernisation on the urban environment are the hardest to mitigate. The sudden demand for housing seen in cities early in the twenty-first century – a period termed by the urbanist Anthony M. Tung 'the century of destruction'[38] – and the obliterating impact of the modern movement on the historic environment have dramatically altered the face of most cities.[39] Inevitably, then, the role of architectural conservation has become increasingly vital and the discipline has won unprecedented popularity.

The political dimensions of conservation are also partly responsible for its rise. As the political theorist Hannah Arendt claimed, 'the reality and reliability of the human world rests primarily on the fact that we are surrounded by things more permanent than the activity by which they were produced, and potentially even more permanent than the lives of their authors'.[40] Simply put, according to Arendt, our memories and our sense of being depend directly on the safeguarding of our built environment. Conversely, and this has repeatedly been accomplished throughout history with deliberate intent, the destruction of the built environment brings about enforced forgetting and disorientation.[41] It is in this sense that the protection of the urban environment through conservation strategies can in fact be used as a tool for memory manipulation, which is what gives conservation a strong political character.

The contemporary interest in conservation is commonly expressed through professional and academic debates worldwide. In a lecture entitled *Preservation/Destruction: OMA – Cronocaos* that was given on 28 March 2011 at the Royal Academy of Arts in London, Ippolito Pestellini and James Westcott of Rem Koolhaas's Office for Metropolitan Architecture (OMA) presented their views regarding the present and future of architectural conservation. In their 'Preservation' exhibition, at the 12th International Venice Architecture Biennale in 2010, the OMA expressed, for the first time, their interest in the past, conservation and history. The projects they showcased addressed two contemporary conflicting tendencies that on the one hand deal with the ambition to preserve progressively larger territories on the planet and, on the other, the desire to obscure any traces of post-war architecture relating to the social project.[42] In the second room of the exhibition, they highlighted the destructive effect conservation has on the linear evolution of time. The overall objective of the OMA's exhibition was to 'document our period of acute CRONOCAOS'.[43] Expanding on this concept, Koolhaas defines *cronocaos* as the contemporary experience of time in urban environments. In his manifesto 'Junkspace', he argues that contemporary cities simultaneously stage the new and the old, the permanent and the temporary. Parts of these cities 'undergo an Alzheimer's-like deterioration as others are upgraded,'[44] and this is a defining aspect of Junkspace. The inhabitants, trapped in these environments, in Junkspace, can no longer distinguish the future, present and past from each other, as they have all merged, creating 'a short circuit in our concept of chronology'.[45]

The OMA exhibition comprises an exemplary illustration of the contemporary interest in conservation and at the same time highlights a rather paradoxical phenomenon. This is the fact that conservation often consents to the eradication of buildings from the post-war period, thus exposing one of its weaknesses, which is precisely its failure to have a fixed agenda or policy for every building.[46] The reasons why post-war architecture often falls outside regimes of conservation are multiple and complex. They span from a lack of advanced technical experience and knowledge in repairing or replacing materials, many of which are modern and often experimental, to insufficient political agreement and public sympathy.[47] In addition to these practical concerns, the conservation of post-war architecture finds many theoretical opponents. As the British architectural writer Owen Hatherley reminds us, the mission of the modern movement was to detach from history, to 'erase the traces',[48] therefore, 'if we want to preserve what remains of Modernism, then we are necessarily conspiring with the very people that have always opposed it: the heritage industries that have so much of Europe in their grip'.[49] As a reaction to the obstacles that the conservation of post-war architecture faces, various international, national and even local organisations have been founded. Characteristic examples are the international work of Do.Co.Mo.Mo. (Documentation and Conservation of Modern Movement) with numerous offices around the globe, the national operation of the C20 (Twentieth Century Society) based in London, or local initiatives like the Hackney Society based in the borough of Hackney in London.

To date, the failure of architectural conservation to establish a fixed, common policy can also be attributed to other parameters, which are equally difficult, if not impossible, to resolve. Current conservation practice is based on what is exceptional and unique. This, however, ignores the significance, or perhaps simply the duty, of preserving something for its mediocrity and ordinariness. Approaching conservation from a global perspective, due to the demands of a globalising culture, exposes further difficulties in finding a common agenda. The Venice Charter for the Conservation and Restoration of Monuments and Sites, for instance, which outlines the internationally approved policy regarding the protection of architectural heritage, is based on Western attitudes, neglecting the fact that some cultures understand permanence and what should be permanent differently.[50] Lastly, there is the evolution of conservation practice: whereas once it dealt with buildings from the distant past, today it also includes buildings from recent years. The 1973 Sydney Opera House in Australia is the youngest building to have made the World Heritage List, in 2007. Similarly, the extension of Museum Liaunig in Neuhaus, Austria, by Querkraft Architekten (fig. 1.3) that opened to the public in 2008 was declared a national monument only five years later, in 2013.

Reflecting on the increasing conservation remit that includes recent buildings, Rem Koolhaas comments: 'From retrospective, preservation will soon become prospective, forced to take decisions for which it is entirely unprepared.'[51] The inspiration for this book therefore is not only the current scale and significance attached to contemporary architectural conservation but also the lack of a consistent policy that accompanies it.

On a philosophical level, architectural conservation faces another major difficulty, as it finds itself opposed to the linear evolution of time. Both conservation and restoration processes manipulate the impact of time on buildings. The former freezes time whereas the latter reverses it so that buildings return to how they looked in past times. These

Figure 1.3 The extension of Museum Liaunig in Neuhaus, Austria, by Querkraft Architekten was declared a national monument in 2013, only five years after it was built. © Querkraft Architekten – Lisa Rastl.

processes can be likened to the philosopher Jeremy Bentham's concept of 'auto-icons'. In one of his last essays, Bentham suggests that after death people's bodies should be preserved with chemicals and displayed in public buildings so that 'every man would be his own monument'.[52] If the concept of the auto-icon is extended beyond the human body to include buildings then conservation and restoration practice can be perceived as having a similar function, which is to protect buildings from maturity and decay so that they become permanent exhibits of their fleeting, temporary selves. This *cronocaos*, the disturbance to the linear evolution of time, raises questions and dilemmas in regard to the criteria behind what is conserved, restored or simply destroyed. In general, architectural conservation faces problems and contradictions that are tricky to deal with and potentially difficult to overcome. The key objective of this book is therefore to explore these issues by revisiting and reassessing long-established key conservation principles, and by developing an alternative theoretical framework that throws light on the current *duty to remember* architecture.

This book approaches architectural conservation through archival theory, a theory broadly used in the emerging academic field of Memory

Studies. This field recognises and promotes the fact that memory, including archives, is studied amongst a number of disciplines. Memory is no longer a subject of enquiry merely for philosophy, psychoanalysis, history and literature, but also for other disciplines such as law, architecture, politics, sociology and anthropology. Memory Studies is essentially a multidisciplinary field, yet more recently academics have started to pursue a more interdisciplinary approach.[53] In this book the concept of memory is approached in an interdisciplinary way with the aspiration to succeed at what the American theorist Marjorie Garber has declared is our task: 'to re-imagine the boundaries of what we have come to believe are disciplines and to have the courage to re-think them'.[54] Architectural conservation is accordingly explored through the fields of architectural history and theory, archival theory, classical mythology, philosophy and psychoanalysis.

It is fair to argue that architectural conservation, and architecture in general, has already been associated with the concept of the archive.[55] Commenting on the architecture of a city, the French philosopher Paul Ricoeur describes how it 'brings together in the same space different ages, offering to our gaze a sedimented history of tastes and cultural forms. The city gives itself as both to be seen and to be read.'[56] In an archive, records are similarly to be seen and to be read; one may draw parallels between the architecture of the city and the archive.[57] A novelty of this book lies in the fact that it does not restrict itself to noting this association but rather makes an active use of archival theory to explore architectural conservation, which agrees with the philosopher Jacques Derrida's ambition for 'a project of a general archiviology … [of] a general and interdisciplinary science of the archive'.[58]

This study of architecture's archival capacity focuses on a single aspect, which is how buildings remember the element of fire. The appropriation of this element lies in the fact that fire is considered by a number of architectural historians to be at the beginning of architectural creation. Yet it is also one of architecture's biggest threats. Fire is arguably present from the birth and until the death of architecture. Hence a parallel thematic thread runs through this book, which, as the title openly suggests, explores the relationship between architecture and fire.

This book is split into seven chapters. Following this introduction, the second chapter starts with an analysis of the concept of the archive. The predominant focus is on Jacques Derrida's psychoanalytic understanding, which has brought about a renewed interest in the study of archives and a radical revaluation of their concept.[59] The second part of this chapter reflects on the fact that architecture is remembered both individually and collectively. Through Henri Bergson's and Maurice Halbwachs' theories of memory it is argued that individual memory on one hand and collective memory on the other can be perceived as operating with the same logic; thus the distinction between the two types of memory can be eliminated. Effectively any reference to memory in architecture essentially implies both types.

Chapter 3 elaborates on fire. It reveals one of its intrinsic qualities, which is its ambivalent nature, which in turn informs the philosophical and methodological approach of this book. I employ the philosophy of the French scientific philosopher Gaston Bachelard, who is known for his extensive work on memory and his psychoanalysis of fire. Yet fire's conflicting nature of temporality dictates the simultaneous employment of an opposing philosophy, and thus I deploy the work of one of Bachelard's main theoretical opponents, Henri Bergson. Accordingly, the contradicting notions of *life and death* and *eternal return* that fire is famously associated with[60] determine the use of the two opposing philosophies: Bachelard's fragmented time and Bergson's temporal continuity. The theoretical conflict inherent in fire also reflects Bachelard's model of knowledge production. This model supports the notion that *knowledge advances dialectically through opposition with existing knowledge*, which is adopted as a methodological tool to investigate the relationship between architecture and memory. Following this methodology, a retrospective account of fire's manifestation in mythology, old rituals, festivals and science demonstrates how the obscure and often confused knowledge that we hold of the element is justified through an archival association.

Chapter 4 discusses architectural conservation. Drawing on archival theory, listed buildings are associated with the function of archives, on the grounds that they both firmly preserve memories from the past. Accordingly, the moment of designation signifies the official archivisation of architecture. Focusing on the English Heritage protection system through Bachelard's and Bergson's theories of time and memory, I argue that archives can be considered as sites of forgetting. As listed buildings function as archives, they can also be considered sites of forgetting: thus the core objective of conservation, which is the perpetuation of memory, collapses.

As the role of conservation appears partially redundant, the following two chapters take a step back and re-evaluate the relationship between architecture and fire. Chapter 5 looks into how architecture evolves by absorbing and accommodating fire on its grounds. Spanning from the flames of the ancient hearth to contemporary architecture, architectural evolution is approached through two perspectives: one examines the way architecture adapts itself to accommodate fire based on the thermal comfort it provides, whereas the other considers fire solely as energy, exploring its expansion in architectural space in energetic terms.

Chapter 6 focuses on a critical moment of architectural evolution, which is the destruction of buildings by fire. It considers the imminent threat imposed by the element and provides a detailed account of how buildings and cities are prepared, or at least should be in principle, to combat potential fire catastrophes. The preparedness to deal with fire incidents is illustrated with a reference to the tragic disaster at Grenfell Tower in London in 2017. Both chapters 5 and 6 share a common objective, which is to assign archival properties to architecture. Both chapters conclude that architecture always carries with it either a reduced or a complete memory of its entire past, making the practice of conservation fundamentally redundant.

In conclusion, the association of architecture with the concept of the archive is approached through Derrida's psychoanalytic interpretation, according to which archives and thus also architecture are approached through the Freudian theory of the drives. This approach suggests that the practice of architectural conservation is an expression of the life drive and a simultaneous repression of the death drive, which leads to a proposal to allow *controlled destruction* to be an integral part of the conservation agenda.

Notes

1. The woman holding the black umbrella is Lilly Jacobs, who is also the assistant editor of the film. The score of *Black Umbrella* was arranged by Hector Castells-Matutano.
2. '1 February 2011: Abandoned Archives and Forgotten Histories Remembered', Close-Up website, accessed 23 August 2018, https://www.closeupfilmcentre.com/film_programmes/2011/abandoned–archives–and–forgotten–histories–remembered/.
3. Frances A. Yates, *The Art of Memory* (London: Pimlico, 2003), 20.
4. Warren Sack, 'Memory' in *Software Studies: A Lexicon,* ed. Matthew Fuller (Cambridge, MA: MIT Press, 2008), 186.
5. Edward S. Casey, *Remembering: A Phenomenological Study*, second edition (Bloomington: Indiana University Press, 2000), 2.
6. Nora proposes that France was probably the first country to experience this obsessive *memorialism*, and he traces its emergence to the mid-1970s. He explains that there are three fundamental occurrences that led to the upsurge in memory during that period. These are 'the after–effects of the economic crisis, the fallout from the post-de Gaulle era, and the exhaustion of the revolutionary idea'. Pierre Nora, 'Reasons for the Current Upsurge in Memory', *Eurozine*, 2002, accessed 19 April 2018, http://www.eurozine.com/articles/2002–04–19–nora–en.html.
7. Pierre Nora, 'Between Memory and History: Les Lieux de Mémoire', translated by Marc Roudebush, *Representations*, 26 (1989), 8.
8. Nora, 'Between Memory and History', 8–9.
9. Nora, 'Reasons for the Current Upsurge in Memory'.
10. In 'Reasons for the Current Upsurge in Memory', Nora explains that there are three types of decolonisation that have taken place in contemporary society and which have led minority

groups to claim their identity. First is an international decolonisation that deals with minority groups that were once colonially oppressed, second is a domestic decolonisation that deals with sexual, religious and social minority groups, and third is an ideological decolonisation that deals with minority groups from ex-totalitarian regimes, which are now free to reunite and revive old, lost, previously banned or destroyed memories and traditions.

11. Anthony Giddens, *The Consequences of Modernity* (Stanford, CA: Stanford University Press, 1990), 64.
12. Ulrich Beck, *Risk Society: Towards a New Modernity* (London: Sage, 1992), 10.
13. Beck, *Risk Society*, 36.
14. Beck, *Risk Society*, 33.
15. Scott Lash and John Urry, *Economies of Signs and Space* (London; Thousand Oaks, CA: Sage, 1994), 10.
16. M. Shanks and C. Witmore, 'Memory Practices and the Archaeological Imagination in Risk Society: Design and Long Term Community' in *Unquiet Pasts: Risk Society, Lived Cultural Heritage, Re–Designing Reflexivity*, ed. Stephanie Koerner and Ian Russell (Burlington, VT: Ashgate, 2010).
17. Rodney Harrison, *Heritage: Critical Approaches* (London: Routledge, 2013), 3.
18. Harrison, *Heritage: Critical Approaches,* 166–7.
19. Jorge Luis Borges, *The Aleph: Including the Prose Fictions from the Maker* (London: Penguin, 2000), 126–7.
20. Borges, *The Aleph*, 131.
21. Borges, *The Aleph*, 129.
22. Borges, *The Aleph*, 131.
23. Friedrich Nietzsche, *On the Uses and Disadvantages of History for Life in Untimely Meditations*, translated by R. J. Hollingdale (Cambridge: Cambridge University Press, 1983), 60–1, quoted in Casey, *Remembering*, 1.
24. *Active forgetfulness* refers to the individual level. Regarding forgetfulness on a bigger scale, in a society for instance, Nietzsche approaches the problem of memory and forgetting through the issue of morality in his second essay *On the Genealogy of Morals*, which carries the subtitle *'Guilt', 'Bad Conscience' and the Like*. His reflections do not analyse directly the obvious side of this subject, the virtues, but rather deal with the dark side of morality, namely guilt and sin. Before his engagement with the problem of guilt (*Schuld*), he concentrates on the relationship developed between humans in relation to their debts (*Schulden*). In a remarkable fashion, he grasps a notion of functional memory residing in the relationship between a debtor and a creditor. Specifically, the creditor's pure interest is to remember the bargain he has offered. Only if he is too generous or soft-hearted can he forget what he is owed. On the other hand, the debtor could easily forget the amount he owes, but the danger of not finding a future creditor or losing his creditworthiness obliges him to remember the promise he has given to repay his debt. In this way, debts are items that are constantly remembered. A similar sense of guilt is entailed within actions of sin. Guilt is an item to be remembered in the mind of the offender and it can only be forgotten after it has been legally established. For Nietzsche, this is the *willed forgetting* that takes place in society alongside *enforced remembering*, both of which want to persuade citizens to abide and conform to laws. See Nietzsche, *On the Uses and Disadvantages of History for Life in Untimely Meditations*, 60–1, quoted in Casey, *Remembering*, 2–3.
25. Sven Spieker, *The Big Archive: Art From Bureaucracy* (Cambridge, MA: MIT Press, 2008), xiii.
26. 'We', Historic England website, accessed 23 August 2018, https://historicengland.org.uk/.
27. Historic England, 'Three Year Corporate Plan 2016–19', May 2016. Accessed 6 February 2019. https://content.historicengland.org.uk/images-books/publications/he-corp-plan-2016–19/three-year-corp-plan-2016–19.pdf/.
28. Peter Brimblecombe and Carlota M. Gross, 'Scientific Research into Architectural Conservation' in *Architectural Conservation: Issues and Developments*, ed. Vincent Shacklock (Shaftesbury: Donhead, 2006), 128.
29. 'Listed Buildings', Historic England website, accessed 23 August 2018, https://historicengland.org.uk/listing/what–is–designation/listed–buildings/.
30. 'Venice and its Lagoon', UNESCO website, accessed 23 August 2018, http://whc.unesco.org/en/list/394.
31. 'Conservation Areas', Historic England website, accessed 23 August 2018, https://historicengland.org.uk/listing/what–is–designation/local/conservation–areas/.

32. According to Article 1 of the 1992 World Heritage Convention, a cultural landscape reflects the 'combined works of nature and of man … They are illustrative of the evolution of human society and settlement over time, under the influence of the physical constraints and/or opportunities presented by their natural environment and of successive social, economic and cultural forces, both external and internal.' 'Cultural Landscapes – UNESCO', UNESCO website, accessed 8 February 2018, http://whc.unesco.org/en/culturallandscape/#1.
33. In June 2009, UNESCO deleted Dresden from the World Heritage List on the grounds that 'a four-lane bridge [was built] in the heart of the cultural landscape which meant that the property failed to keep its "outstanding universal value as inscribed".' 'Dresden is Deleted from UNESCOS's World Heritage List', UNESCO website, accessed 8 February 2018, http://whc.unesco.org/en/news/522.
34. Jorge Otero-Pailos, 'Editorial: Chance Architecture', *Future Anterior* 3: 2 (Winter 2006), v.
35. James Marston Fitch, *Historic Preservation: Curatorial Management of the Built World* (Charlottesville: University Press of Virginia, 2001), x.
36. John H. Stubbs, *Time Honored: A Global View of Architectural Conservation. Parameters, Theory and Evolution of an Ethos* (Hoboken, NJ: Wiley, 2009), 54.
37. Stubbs, *Time Honored*, 98.
38. Anthony M. Tung, *Preserving the World's Great Cities: The Destruction and Renewal of the Historic Metropolis* (New York: Clarkson-Potter, 2001).
39. Stubbs, *Time Honored*, 99.
40. Hannah Arendt, *The Human Condition* (1958; repr. Chicago: University of Chicago Press, 1998), 95–6.
41. Robert Bevan, *The Destruction of Memory: Architecture at War* (London: Reaktion, 2006), 8.
42. Rem Koolhaas, 'Cronocaos', *Log*, 21 (Winter 2011), 119.
43. Koolhaas, 'Cronocaos', 119.
44. Rem Koolhaas, 'Junkspace', *October* 100 (Spring 2002), 180.
45. 'The Cronocaos Exhibit at the New Museum: Rem Koolhaas Says Make no Little Plans', *UrbDeZine*, 2012, accessed 18 August 2018, https://urbdezine.com/2011/08/10/the–cronocaos–exhibit–at–the–new–museum/.
46. The lack of a consistent conservation policy applies also to postmodernist architecture, which is surprising as postmodernism itself echoes the past.
47. Susan Macdonald, Kyle C. Normandin and Bob Kindred, *Conservation of Modern Architecture* (Shaftesbury: Donhead, 2007), 1–2.
48. Owen Hatherley, *Militant Modernism* (Winchester; Washington, DC: O Books, 2008), 3.
49. Hatherley, *Militant Modernism*, 5.
50. Seung-Jin Chung, 'East Asian Values in Historic Conservation', *Journal of Architectural Conservation: The International Journal for Historic Buildings, Monuments and Places*, 11: 1 (March 2005), 55–6.
51. Koolhaas, 'Cronocaos', 122.
52. Jeremy Bentham, *Auto-Icon; or, Farther Uses of the Dead to the Living: A Fragment* (London, 1842), 3–4.
53. H. L. Roediger and J. V. Wertsch, 'Creating a New Discipline of Memory Studies', *Memory Studies* 1:1 (January 2008), 9.
54. Marjorie Garber quoted in Marlene Manoff, 'Theories of the Archive from Across the Disciplines', *Libraries and the Academy* 4:1 (2004), 96.
55. Mark Wigley, 'Unleashing the Archive', *Future Anterior* 2:2 (Winter 2005), 12.
56. Paul Ricoeur, *Memory, History, Forgetting* (Chicago: University of Chicago Press, 2004), 151.
57. Commenting on the parallelism between buildings and archives, Kent Kleinman in *Archiving/Architecture* develops an alternative argument by distinguishing two modes of architectural existence: the built artefact and the representations of it, i.e. the drawings, models, and visualisations produced by the architect. For Kleinman, the representations of architecture comprise the true and original architectural archive, as they reflect the pure aspirations of the creator. On the other hand, because the built artefact is subject to changes caused by weather, ageing, inhabitation, and so on, it faces issues of originality and authorship and thus it cannot be perceived as solid and complete as the representation of it. Moreover, based on Roland Barthes's *The Death of the Author*, Kleinman draws parallels

between a building and a document, saying that the former is 'a record every bit as open to, and in need of, interpretation as is a document housed in an archive'. Kent Kleinman, 'Archiving/Architecture', *Archival Science* 1 (2001), 324.

58. Jacques Derrida, *Archive Fever: A Freudian Impression*, trans. Eric Prenowitz (Chicago: University of Chicago Press, 1996), 34.
59. Jennie Hill, ed., *The Future of Archives and Recordkeeping: A Reader* (London: Facet, 2010), 8–9.
60. Gaston Bachelard, *The Psychoanalysis of Fire*, translated by Alan C. M. Ross with preface by Northrop Frye (Boston: Beacon Press, 1968), 7.

2
On Archives

> They are planning a 'City' for Moscow in the Nikolaskaya, Varvarka and Ilinka districts, corresponding to London's belly of capitalism. In the suburbs they are building 'pleasant estates' and suchlike for the workers. Where are such paltry Utopias born? In the archives!
> When they came to design a new Kemenny Bridge over the Moska River for their projected Utopia ... they dispatched a gravedigger to 'carry out a thorough excavation in the archives, to unearth a historical reference to the Kamenny bridge' ... Now we have had the opportunity of becoming acquainted with the forms resulting from this search in the archives ....
>
> El Lissitzky, *The Catastrophe of Architecture* (1921)

> Nothing is less reliable, nothing is less clear today than the word 'archive'.
>
> Jacques Derrida, *Archive Fever: A Freudian Impression* (1995)

In the dialogue between Socrates and Phaedrus that is delivered in Plato's *Phaedrus* (*c*. 370 BC), the former narrates the myth of the Egyptian god Theuth, the discoverer of writing. Theuth enthusiastically approaches the Egyptian king Thamus to share his discovery and to convince him that everyone in Egypt should learn how to write. Theuth justifies his discovery as a *pharmakon* (potion) of memory and wisdom. In return Thamus argues that writing would lead to oblivion. Writing for Thamus would replace the function of memory, as people would depend exclusively on signs, which are external; thus the result would be merely the appearance of wisdom rather than the real thing. In the narration of this myth an ambiguity arises in the meaning of the word *pharmakon*. According to Derrida, *pharmakon* could imply either a remedy or a poison, and this ambivalence is not accidental. Derrida argues that Plato's discourse is consistent with this sort of ambiguity. Thamus's response could equally imply the opposite, which is that *pharmakon* is also a remedy.[1]

The ancient ambiguity surrounding writing, intentionally invoked by Plato, can equally be traced in the function of memory. As seen in Plato's *Theaetetus*, memory operates by inscribing or imprinting the Divine Ideas upon the wax of the soul, though in this case writing is treated as a remedy. Commenting on the ambiguity produced through the use of the word *pharmakon*, the French philosopher Paul Ricoeur notes that 'Plato's *Phaedrus* [is] the mythic birth of the writing of history … must we not ask whether the writing of history, too, is a remedy or a poison?'[2] Archives, the places where written records or the writings of history are sheltered, are similarly coloured by this ambiguity, as they carry within their very substance the notion of *pharmakon* as either a remedy or a poison.

Archives are most commonly perceived as spaces destined for the storage and preservation of records. These records are arranged following the rules of an elaborate classification system, and they are left in a state of stasis until someone accesses them for research purposes. The processes that take place within an archive are complex and sophisticated, faithfully reflecting the narrative about the multiple operations unfolding silently in the Bibliothèque Nationale de France in Paris in Alain Resnais's short film *Toute la mémoire du monde* (1957). Referring to a book in this library that has been suddenly chosen from amongst other records, Jacques Dumesnil, the narrator in Resnais's film, comments descriptively:

> Before, it was part of a universal, abstract, indifferent memory where all books were equal and together basked in attention as tenderly distant as that shown by God to men. Here it's been picked out, preferred over others. Here it's indispensable to its reader, torn from its galaxy to feed these paper-crunching pseudo-insects, irreparably different from true insects in that each is bound to its own distinct concern.[3]

The selection of records is usually intentional and follows a rational searching process through the long catalogues of the archive. Yet it is not uncommon that records are accessed unintentionally, encountered through a process of serendipity. Likening this to an archival flâneury, the sociologist Mike Featherstone remembers the unorthodox searching methods of the historians Norbert Elias and Michel Foucault, who accessed documents randomly '"on the diagonal" … so that the unusual juxtapositions they arrived at summoned up new lines of thought and possibilities to radically re-think and reclassify received wisdom'.[4] Archives are therefore sites of potential discoveries where the work of the researcher resembles that of the archaeologist. Both *dig into* unmapped territories, searching for historical clues and interpreting them for credible glimpses of the past.

Reflecting on the digging activities of researchers and archaeologists, the Austrian artist Lois Weinberger employed an innovative method that involved literal archaeological works. He chose the attic of his family house in Stams, Austria, as the site for his research[5] and closely examined the accumulated waste found under the floorboards. Through an exhaustive sifting of every piece of waste, he categorised 2000 objects dating from the fourteenth to the nineteenth century. Every object that was brought to the surface became a component of *Debris Field*, an installation of found objects, drawings, photographs and notes (fig. 2.1) that was presented in the *Documenta 14* exhibition at EMST, the National Museum of Contemporary Arts, in Athens in 2017.

In this installation, Weinberger's attic is an archive, a site of potential discoveries, and the excavated objects form the records of the archive, which await the arrival of the artist/archaeologist to assess their historical significance in the present.

Today, a time of boundless archiving, this task of interpreting and judging the archive's content is profoundly significant. What material is important or relevant to somebody's research? Is there an end to this research? Does the material found in an archive reflect the truth? These are just some of the dilemmas one might come across when browsing through records in an archive. The same confusion with regard to archives

Figure 2.1 Lois Weinberger's installation *Debris Field* at EMST, National Museum of Contemporary Arts, Athens in 2017.
© Studio Weinberger – Mathias Völzke.

is expressed through Ilya Kabakov's character in the installation *The Man Who Never Threw Anything Away* (1988).[6] Throughout his life, the character collects useless objects and hangs them from the ceiling with a string. At the end of the string,[7] a small piece of paper explains where the object was found, in an attempt to connect it back to his memory.

> [T]his is the memory associated with all the events connected to each of these papers. To deprive ourselves of these paper symbols and testimonies is to deprive ourselves somewhat of our memories. In our memory everything becomes equally valuable and significant. All points of our recollections are tied to one another. They form chains and connections in our memory, which ultimately comprise the story of life.[8]

Later in the story however, he finds himself amongst all these useless objects and exhausted questions:

> Does the dump and its image summon my imagination over and over again, why do I always return to it? Because I feel that man, living in our region, is simply suffocating in his own life among the garbage since there is nowhere to take it, nowhere to sweep it out – we have lost the border between garbage and non-garbage space.[9]

The confusion of Kabakov's character eloquently reflects the ambiguity attached to archives, which, either as a remedy or a poison, can be beneficial or problematic respectively.

The perception of archives as static depositories of ordered records has prevailed since Hilary Jenkinson's publication of *A Manual of Archive Administration* in 1922. For Jenkinson, archives 'themselves state no opinion, voice no conjecture; they are simply written memorials, authenticated by the fact of their official preservation, of events which actually occurred and of which they themselves formed a part … [and provide] an exact statement of facts'.[10] Archives are understood as passive, disinterested and fixed entities that reveal *one truth*. The temporal dimension of their existence is unique and linear, as they rise from stillness to movement until they return back to their original place for storage. More recent debates on archival theory, however, disregard the idea of a sole truth behind archives. In the light of postmodernism, the world cannot be adequately explained and justified in a single, unified approach; thus this *one truth* must become 'a series of contingent truths'.[11] Roland Barthes's *Death of the Author*, for instance, accepts that the author's intention is detached as soon as writing begins, and comprises one of the first post-modern texts that clearly confront Jenkinson's *one truth*. Yet the most influential person to determine the future of archives is the French philosopher Jacques Derrida.[12]

The Derridian archive

In *Archive Fever*, Derrida produced a critical and intense study of the notion of the archive drawing on Freudian psychoanalysis. *Archive Fever* is a translation from the French of a lecture that Derrida gave at an international colloquium called *Memory: The Question of the Archives* that took place at the Freud Archives in 1994 in London. In this book, Derrida revisits topics that he had long been preoccupied with, especially in *The Post Card*. That book, published in 1980, starts with *Envois*, which narrates the story of Derrida on a day in Oxford accompanied by Jonathan Culler and Cynthia. On that day Derrida enters a bookshop where he comes across a postcard with an image from Matthew Paris's thirteenth-century fortune-telling book. This postcard depicts an image of plato, whose name is spelled with a small *p*, and of Socrates. Being extremely impressed by this postcard, Derrida buys every copy the shop holds and starts sending them to a person very close to him. During the reading of *Envois*, the reader gradually understands that these postcards are sent to a woman Derrida is in love with, and the postcards are thus in essence love letters. Derrida recommends that the reader considers these postcards 'as the remainders of a recently destroyed correspondence. Destroyed by fire or by that which figuratively takes its place, more certain of leaving nothing out of the reach of what I call the tongue of fire, not even the cinders there are [*s'il y a là cendre*]'.[13] Derrida suggests that these postcards/love letters comprise a type of archive that carries the memory of a broken-up correspondence, which appears to be of a confused and heated nature. On 6 June 1977, a section of an envoi reads:

> If you had listened to me, you would have burned everything, and nothing would have arrived. I mean on the contrary that something ineffaceable would have arrived, instead of this bottomless misery in which we are dying. But it is unjust to say that you did not listen to me, you listened closely to the other voice (we were already a crowd in that first envelope) which asked you not to burn, to burn in order to save. Nothing has arrived because you wanted to preserve

> (and therefore to lose), which in effect formed the sense of the order coming from behind my voice, you remember, so many years ago, in my first 'true' letter: 'burn everything'. You had answered me the next day, and this is how your letter ended: "] 'The letter ends on the exigency of this supreme pleasure: the desire to be torn by you' ... I am burning. I have the stupid impression of being faithful to you ... I am waking up. I remember the ashes. What a chance to burn, yes yes.[14]

Throughout *Envois* Derrida exposes a peculiar aspect of archives which he expands upon a few years later in *Archive Fever*. This aspect deals with a contradiction inherent in them according to which memory is not only preserved in archives but also burnt and destroyed. By their nature therefore archives can be considered as carrying this trauma, which gives one a feverish compulsion, an *archive fever*, to look into and find the origin of memory.[15] Commenting on the relationship between *The Post Card* and *Archive Fever*, Herman Rapaport describes the former as the 'phantom limb, something important that has been cut off and that haunts'.[16] The difference between them is that *The Post Card* deals with an archive of a private correspondence, between lovers, whereas *Archive Fever* deals with Freud's public archives.

Following *The Post Card* and having traced the problem of an inherent contradiction in archives, Derrida elaborates his thoughts further and develops a consistent theory in *Archive Fever*. In the opening note of *Archive Fever*, Derrida discusses the etymology of the word archive, referring to the Greek word *arkhe* (αρχή). This word has two meanings, *commencement* and *commandment*. Commencement refers to the origin, or as Derrida says, it has sequential implications, whereas commandment, or giving orders, refers to the law. Both commencement and commandment take place simultaneously under one word, *arkhe*. The word archive derives from the Greek *arkheion*, which refers to the building where the *arkhons* resided. Part of the *arkhons*' political power was to give orders, to command, and it was at their residences, in the *arkheia*, where documents were kept safe and in order.[17] It is thus through the very physicality of the archive, that the archive itself comes '[t]o shelter itself and, sheltered, to conceal itself'.[18] Derrida approaches the notion of the archives from a novel perspective and deconstructs a central function of it, which is the one of inscription. Initially, he discusses the entirety of Freud's work on psychoanalysis as an archive of his legacy. During his career as a psychoanalyst, and even after his death, Freud's archive provided ongoing evidence that could be used to defend his reputation.

Derrida also comments on printing processes that take place in archives and associates them with Freud's 'Mystic Writing Pad' (*Wunderblock*) and the unconscious.

In the 1926 paper 'A Note about the "Mystic Writing Pad"' Freud draws parallels between our perceptual apparatus and the Mystic Writing Pad (*Wunderblock*). He explains that the *Wunderblock* is a small apparatus, which is different to paper and slate in the sense that it 'can really deliver both components: an always ready receptive surface and lasting traces of the notations made'.[19] For Freud, the function of the perceptual apparatus involves an internal printing process on some external, virtual substrate, or as Derrida explains, the perceptual apparatus 'integrates the necessity, inside the *psyche* itself, of a certain outside'.[20] Commenting on this function, Derrida adds that 'the theory of psychoanalysis is not only a theory of memory but also a theory of the archive'.[21] Derrida does not restrict himself only to this archival understanding of Freudian memory[22] but also examines private inscription. Specifically he discusses Freud's circumcision, as an inseparable aspect of his Jewish inheritance, and a gift that Freud was given by his father Jakob on his thirty-fifth birthday in Vienna in 1891. This gift was the Bible that Freud studied as a child with a new leather binding. Derrida explains that Freud's circumcision and his father's gift comprise the original archive, as they carry the mark that is both exterior to him and inscribed on his body and thus permanently with him.

One of the characteristics of archives, Derrida explains, is violence. Violence occurs as soon as something is archived, for this act implies the establishment of a law and also its enforcement, which is the same as 'the violence of power (*Gewalt*) which at once posits and conserves the law'.[23] An archive is thus the place where power is exercised. A visit to a library, for instance, demands acquaintance with a certain protocol. One needs to know how books are categorised and where they are stored, what books can be borrowed and what books are for reference only. Some parts of the building may not be accessible to the public and others may have limited accessibility. A visit to the library will reflect the archive's institutional function, its law. The moment of preserving the past is violent in the sense that certain aspects relevant to the archive are only then revealed. The archive decides what is worth remembering.[24] And this selection process automatically determines what is to be forgotten. On the one hand therefore, this violence can be perceived as positive, as it opens up the ground, broadens the boundaries and gives freedom to the archived material, to memory. On the other hand, the same violence can be perceived as negative, as it restricts memory by ignoring part of it. This is what Derrida refers to as an *archontic* quality of the archive, which is always followed by the *anarchontic*.[25] Derrida makes a further association between archives and political power, reminiscent of Michel Foucault's understanding of the archive,[26] claiming that '[t]here is no political power without control of the archive, if not of memory. Effective democratisation can always be measured by this essential criterion: the participation in and the access to the archive, its constitution, and its interpretation.'[27]

In the earlier reference to *The Post Card*, I mentioned one of the archive's intrinsic qualities, which is archive fever. Derrida elaborates further on this quality in *Archive Fever*: 'It is to have a compulsive, repetitive, and nostalgic desire for the archive, an irrepressible desire to return to the origin, a homesickness, a nostalgia for the return to the most archaic place of absolute commencement.'[28] He draws parallels between this desire to return to the origin, to the primordial memory, and the Freudian death drive, a focal point in Freud's psychoanalytic theory that is often neglected or disputed.[29]

The Freudian death drive

Freud introduces his theory of the death drive in *Beyond the Pleasure Principle,* his most philosophical essay, written in 1920.[30] This essay examines the reasons why the mind impels us towards unpleasant and often painful experiences. He focuses on the dreams of shell-shocked veterans from World War I who suffered from traumatic neurosis.[31] Their dreams were 'characterised by their reintroducing the patient again and again into the situation of his accident, a situation from which he awakens with renewed fright ... fixations ... familiar to us in hysteria'.[32] Freud's main interest is why these dreams repeat themselves on a daily basis and he assumes that they 'seek to master the stimulus by developing anxiety, the lack of which was the cause of traumatic neurosis'.[33] Through the re-enactment of a traumatic event in dreams, he explains, the mind tries to create the anxiety that was lacking during the original event. During this process, the mind ignores the emotional stability and wellbeing of the person in the present and goes beyond pleasure, to the core of the traumatic event. This compulsion to repeat a traumatic event, known as a 'repetition compulsion',[34] has an inherent quality always to point towards the past, which is something that fascinated Freud as 'new and radical'[35] and applies to all drives in general. Freud remarks that, whereas drives are commonly perceived as urges that point towards change and development, in actual fact they do the opposite. Organic drives point towards the restoration of previous states: thus they are conservative and regressive.[36] Commenting on the nature of organic drives, Freud adds that if 'every living being dies for *internal* reasons, returning to the inorganic, then we can only say that *the goal of all life is death*, and, looking backwards, that *the non-living existed before the living*'.[37] The drive that pushes a person towards extinction, towards an inanimate, inorganic state, is therefore the death drive (*thanatos*), which, through repetition compulsions, appears in opposition to but also in balance with the life drive (*eros*), the drive that is concerned with the creation and preservation of life. The life drive is topographically different from the death drive. The former is situated within living matter, whereas the latter operates silently from outside.[38]

When Freud first introduced the concept of the death drive in *Beyond the Pleasure Principle*, he had doubts regarding its theoretical validity as a 'false profundity or mysticism'[39] but later, in *Civilisation and its Discontents*, he admits: '[my views of the death drive] have taken hold of me so strongly that I can no longer think in any other way'.[40] Based on the theoretical hypothesis of the death drive, Freud proceeds further, giving an explanation for certain clinical illnesses. He argues that the death drive can be expressed in two ways: one that is outwards as a sadistic aggression and another that is internal and expressed as masochism. He further suggests that the necessity for externalising aggression is imperative, as the same aggression can be directed inwards and cause self-destruction.[41] This implies that sadistic behaviour is normal and also beneficial for the organism.[42] Returning to Derrida, the archive fever, or else the desire to return to the first memory, to the absolute beginning, is fundamentally linked to the Freudian death drive that points towards an inorganic state, towards the destruction of memory. This association implies that the death drive, through repetition compulsions, operates as a self-contradictory force that simultaneously generates and destroys the archive.

Archives and the city

The concept of the archive can extend beyond the physical boundaries defined by its walls and can include an entire city. This thought is shared with the British historian Patrick Joyce who, responding to Derrida's statement that the archive is a place where power is exercised, makes a fascinating enquiry into how archives are adjusted to accommodate changes that take place on national and social levels. In his essay 'The Politics of the Liberal Archive', he focuses on the key role of the public library and the way it embodies and reflects the idea of the public in nineteenth-century liberalism. He explains that the first truly public library appeared in the United Kingdom only after the Library Act of 1850 was enforced. The Public Record Office founded in 1938 or the British Museum's Reading Room established in 1753 cannot be considered public archives due to their limited access.[43] Following the Library Act of 1850, the concept of the public library was redefined: the public library was for the first time open to everybody, to the *demos*, and its mission was to 'civilise the working class by giving this class access to the public realm'.[44] Nevertheless the concept of the public library expanded further during the post-war period when an 'accessible network of well-stocked public libraries was celebrated as an important buffer against totalitarianism – a vital organ of democracy, which exemplified Britain's essential historic traditions of individual liberty and social empowerment'.[45]

Joyce makes a noteworthy observation on the centrality of the library in nineteenth-century liberalism. Following the 1850 Act, the organisation of libraries started to develop an interest in the *centre*, and this has various interpretations. Firstly public libraries followed the example of the British Museum Library, applying a balanced cultural policy between the capital, that is the centre, and the rest of the country, that is between the local and the national. In terms of the internal design and layout of libraries, there was a change too. Libraries adopted the concept of Bentham's *Panopticon* and thus the public was supervised by a centrally located member of staff.[46] The British Museum Reading Room built in 1857 features a centrally raised platform from which the staff *policed the public*. In this way, the notion of the Central Library comes to reveal 'the local dimensions of liberalism'.[47] Further, central libraries occupy the physical centre of towns and cities to ensure equal accessibility to all patrons. Closing his essay, Joyce expands further on the concept of the archive, claiming that in essence it embodies something broader and more diverse, which is 'the street and the built environment',[48] the city itself.

Joyce's understanding of the archive, which is not spatially restricted within the walls of the library but includes the whole street, the city and its everyday life as a repository of memories, is not novel. In literature, Marcel Proust's *In Search of Lost Time* (*A la Recherche du temps perdu*) is a powerful attempt to 'recapture through memory traces, the richness of that everyday life'.[49] Equal to Proust is Walter Benjamin's famous *Arcades*

Project (*Passagenwerk*) in which he narrates his memories of flânerie[50] in nineteenth-century Paris, or in his essay 'Naples' (1924), written with his lover Asja Lacis.[51] For Proust and Benjamin, both the city and its everyday life function as archives. They provide material to be retrieved and then carefully assembled into textual form.

Archives, historiography and monumentality

In her discussion of archives, the architectural historian Beatriz Colomina reflects on the production of history. She perceives archives as fragmented entities that hide in a private and 'messy space'.[52] The task of the historiographer, she explains, is to bring these entities *out* of this private, disordered world and expose them to the public in a carefully assembled way. This orderly displacement of archival material from the private inside to the public outside is what the writing of history comprises.

Similarly, for Paul Ricoeur, the archive signifies the moment when the historiographical process is written, following the stage of the testimony. The testimony, for him, opens up the historiographical process and archives come to capture it. In *Memory, History, Forgetting*, Ricoeur ponders that if writings are the main material held in a library and written testimonies the main collection of it, then everything can be archived. He has doubts, however, about the transition between the first two stages of the historiographical process, from testimony to archive, or from oral to written history. Such doubt relates to the transition from a living memory to a written record, which, like a *pharmakon*, can be either poison or remedy. In another essay, Ricoeur remembers the French historian Jacques Le Goff, who asserted that in past centuries archives were associated with *monuments*, such as the *Monumenta Germaniae Historica* of 1826. He explains that monuments, which are in most cases created by the state, express an ideology and embody collective memory. Archives, on the other hand, which are associated with documents, have different intentions; they are independent of ideology and therefore more subjective. For Ricoeur, however, the association of archives with the notion of the document was never really made in this sense. For him documents 'attack against the conditions of historical production and its concealed or unconscious intentions … we must say with Le Goff that once its apparent meaning is demystified, "the document is a monument."'[53]

Ricoeur's equation of the document with the monument is reminiscent of Theodor Adorno who, reflecting on Paul Valéry's statement that in museums we 'put the art of the past to death,'[54] advocates provocatively that the museum and mausoleum are not just phonetically linked. He suggests that a museum, which is an archive in a broader sense, is like a mausoleum, a monument built with the intention to convey a specific meaning and ideology.[55]

Digital and dynamic archives

Contemporary archival discourse does not only scrutinise traditional archives, namely the physical documents found in libraries and the collectible objects displayed in museums, but also a more recent type of archives, the digital ones. Digital archives refers to magnetic storage media (CDs, DVDs, hard drives etc.) that have the ability to store data effortlessly and quickly. Today information technologies are so advanced that digital archiving is possible and easily accessible, to the extent that 'we [all] are miniarchivists ourselves'.[56] According to Mike Featherstone, 'to be is to record',[57] whereas for Adrian Mackenzie 'to die is to be disconnected from access to the archives, not jacked-in or not in real time'.[58] The data stored in digital archives is encoded in the form of 1s and 0s, and can be transmitted through network channels locally, such as a home or office computer network, or globally through the Internet. The Internet itself is not an archive in the traditional sense. As Jussi Parikka explains, the Internet is a *dynamic archive* that is open-ended and perpetually updating itself in real life.[59] The distribution of digital data around the globe takes place instantly. This instantaneity has a direct impact on the traditional understanding of knowledge dissemination. The physicality of the traditional, static archive has dissolved and the centrality attached to it is somewhat undermined. Instead, digital archives are fluid, dynamic and decentred.

Digital archives appear at first to overcome the fragility of traditional ones,[60] yet their increasing use raises concerns of a different nature. Digital archives are susceptible to invasion, erasure and overwriting. Their protection is a matter of national security with data centres being fortified and often hidden away in remote locations, like the Green Mountain data centre in Norway that is embedded in the rock of a mountain or the CyberBunker in the Netherlands that is located in a former nuclear bunker. Digital archives are commonly equipped with anti-malware software to avoid remote hacking activities that can erase or overwrite their data.

Individual memory: An archival reading

An archival process unfolds in three stages. The initial stage is the writing of archival material. The second stage deals with the storage of this material and the third stage is the stage of retrieval. This three-faceted operation of archives – writing, storing, retrieving – is to be found in the function of memory too. This is confirmed by Pierre Nora, who suggests that '[m]odern memory is first of all archival. It relies entirely on the specificity of the trace, the materiality of the vestige, the concreteness of the recording, the visibility of the image.'[61]

The archival understanding of memory follows a long tradition in Western thought, which is usually approached from the point of the individual. This tradition stretches back to the dawn of philosophy with the introduction of the Platonic universe of ideas in the fifth century BC. In *Theaetetus,* Plato likens memory to a block of wax that exists in our souls. In processes of perception through our senses, we imprint our thoughts and perceptions on this block of wax. Knowledge is something already possessed. There are forms of ideas in our memories, which the soul has been acquainted with since time immemorial, yet everything is forgotten as soon as one is born. Birth indicates forgetting. True knowledge is achieved by fitting the imprints from sense impressions, which themselves are reflections of a higher reality, onto the prime mould or imprint of this reality. Knowledge can thus be re-attained from within.[62] In *Phaedrus*, Plato sets out to convince his readers that all knowledge is an attempt to remember the things that all souls once witnessed. Attaining knowledge is therefore remembering something forgotten, which gives this inquiry a recollective character (*anamnesis*).[63]

Later in the history of philosophy, Aristotle (384–322 BC) offers a treatise that brings about the complete secularisation of memory, yet memory still appears to operate archivally. His theory of memory and reminiscence is developed in his work *On the Soul* (*De Anima*), according to which imagination is initially aroused by the five senses. Imagination acts as a catalyst between the stage of perception and the formation of thought. Perceptions are manifested in the form of images, and all knowledge that one acquires stems from sense impressions which are functions occurring in the soul. Aristotle denies the possibility of human thought without the employment of mental images (*phantasia*).[64] In *On Memory and Reminiscence* (*De Memoria et Remininiscentia*), he further relates imagination to memory. The difference between them is traced to the element of time. Whereas imagination is a sense impression of things in the present, memory deals with sense impressions obtained in the past. Aristotle likens mental images to paintings and, in a similar fashion to Plato, the formation of mental images is perceived as a movement, like the movement of creating a seal on wax. Regarding the differences between memory and recollection, Aristotle states that recollection is the recovery or recapture of sense impressions that one perceived in the past and that this is a conscious effort on the part of the inquirer.[65]

In ancient Rome memory was perceived as an art. It was called mnemonics and mainly practised by trained orators. This art of memorisation did not simply lead to a blind recitation of a prefabricated speech. Instead, orators had the ability to assemble their speech and deliver it eloquently *ex tempore*. Mnemonics was therefore not a static art but rather a compositional one, which relates to what we call today imagination and creativity. This art of memory is a spatial art. Memorisation is accomplished through the placement and storing of mental images (*imagines*) within certain mental places (*loci*) that were very familiar to the artist, such as remembered versions of his house or part of a city. The retrieval of a stored memory image is a matter of strolling within this mentally constructed building or landscape in an attempt to locate this image. Since memory images hold distinct positions within this construction, their recollection can be delivered without forgetting and in the correct order. Mnemonics was practised up until the early modern period, after which there was a notable change.[66]

During the early modern period, memory was not solely conceived as a blind reproduction of a memory from the past, it was also coloured by the individual's engagement in the process. John Locke, for instance, introduced a new concept of memory based on empiricism, according to which remembering is highly affected by the interests of the inquirer. Memory, for Locke, develops through an accumulative process. In *An Essay Concerning Human Understanding*, Locke draws parallels between an infant's mind and a white piece of paper that is empty of characters and ideas (*tabula rasa*). Memories are imprinted on this blank paper (the mind) through empirical processes, during the individual's life.[67] In this sense memory is still conceived of as following an archival process.

The archival function of memory is equally predominant in discourses of late modernity. Freud's psychoanalytic work on memory is reminiscent of the Platonic model. It relates both to the Socratic method of accessing knowledge, an innate recollection attained by the soul of the inquirer through a dialectical process, and to the Platonic model, where in unconscious records memory is not stored in a physical part of the brain or elsewhere in the body – instead it is registered in consciousness through external stimuli and eventually imprinted in the unconscious, like the imprints on Plato's block of wax.[68]

Collective and cultural memory: An archival reading

The first person to coin the term collective memory was the French philosopher and sociologist Maurice Halbwachs. The Durkheimian school of thought evidently influenced Halbwachs, as he understood memory as a social phenomenon. His theory of collective memory also carries resonances of Henri Bergson's philosophy, one of his early mentors, whose theory of memory is approached from an individual's point of view.

Bergson intended to give new dimensions to the metaphysics of memory, and his project opened up new horizons in contemporary thought. One of his most innovative contributions was his attempt to establish a philosophical worldview that reflected the climate created by Einstein's *Theory of Relativity*.[69] Philosophical thought until then had been preoccupied with space rather than time, for space was regarded as a conceivable prerequisite inherently native to our senses. Bergson considers that there are two main types of memory: *habit memory* and *representational memory*. Habit memory refers to the automatic behaviour of the body acquired through repetition. As with forming a habit, one has to deconstruct an action into parts, memorise them individually through repetition, and then reconstruct all parts in a whole and put it into action. In the instance of learning a text by heart, the memory of each reading constructs an individual image. When the text is finally memorised, it emerges as a composite of all previous recollections, which is the habit memory. The other type of memory, representational memory, refers to a function that stores all personal experiences to something external. Representational memory, Bergson explains, 'records, in the form of memory-images, all the events of our daily life as they occur in time; it neglects no detail; it leaves to each fact, to each gesture, its place and date'.[70] As opposed to habit memory, representational memory does not have a utilitarian character and it manifests itself spontaneously.

Halbwachs' theory of memory, which was formulated as a reaction against psychological treatises of that period that treated humans as isolated units, depends upon the social environment in which it is created. For him, society is the medium through which memories are made. Recollection takes place in society, through which memories obtain a locale. Whereas for Bergson recollection occurs as a mental leap of the individual in the past, for Halbwachs it demands a leap in the social framework of the society.

Halbwachs' book *On Collective Memory* begins with a chapter on dreams. Dreams are composed of random images containing mixed-up memories that do not correspond to the essence of real memory.

Dreams are different, as they do not relate to a social context. This approach is diametrically opposite to Bergson's approach, in which representational memory finds its ultimate manifestation in dreams. As Anne Whitehead notes, Halbwachs' detachment of memory and dream implies that '"individual memory" ... represents ... a paradoxical formulation, because memory itself collapses once we enter into a state of isolation'.[71]

The past according to Halbwachs is not preserved in a sense that it can recur, but it is reconstructed in the present. In this sense, as memories of the past refer to social contexts of previous times, they come to appear different in the present, lacking their original essence. He illustrates this argument with a parallelism between past memories and the stones of Roman houses, according to which 'their antiquity cannot be established by their form or their appearance but only by the fact that they still show the effaced vestiges of old character'.[72] So, as the reconstruction of past memories takes place in the present, the present social framework affects the process of recollection too. In a later discussion, he comments that memories can be classified in different periods of our lives not because they happened to occur in these specific periods, but because at that time we participated in certain *social groups*. Effectively there is no difference between recent and old memories, because recollection relates to a particular social framework as opposed to a specific time. A social group can be the family, school, work, social class, religion etc., and each of these groups has particular memories that belong to them alone. Life is an open repository of memories that the mind reproduces constantly in the present. Through this ongoing process of accumulating memories, 'a continual relationship, a sense of our identity is perpetuated'.[73] Regarding the elderly and their gradual failure to remember certain memories from the past, Halbwachs argues that this is because the social group to which these memories belong starts to disappear. As members of this social group die, the memories that are associated with it die too. In this respect Halbwachs' idea of collective memory relates to the span of a social group's living memory. After that, history comes to replace collective memory.

Reflecting on the fact that collective memory fades as the members of a social group disappear, Paul Connerton in *How Societies Remember* questions how collective memory is transmitted from generation to generation. To answer this question, he focuses on what Halbwachs fails to clarify, which in turn reflects the common ground between Halbwachs' theory of memory and Bergson's. Connerton explains that images and knowledge from the past are transferred to the present through the practice of commemorative ceremonies and bodily practices. The repetition of commemorative ceremonies, as forms of rituals, establishes continuity with the past. What is crucial is not the form of the ritual but the content it carries. In effect the repetition of a ritual in the present leads to the re-enactment of actions that are rooted in the past. Rituals transmit knowledge from the past through specific language, postures, gestures and movements that take place during their practice. In general, bodily practices carry habitual memories within their very substance. As Connerton claims, '[m]any forms of habitual skilled remembering illustrate a keeping of the past in mind that, without ever adverting to its historical origin, nevertheless re-enacts the past in our present conduct. In habitual memory the past is, as it were, sedimented in the body.'[74] Therefore the conceptual gap between Halbwachs' collective memory and its ability to pass from generation to generation can be filled in with Connerton's clarification, which in essence reflects the Bergsonian model of habit memory. For this reason, as Whitehead points out, the notion of collective memory cannot stand alone, independent of the long tradition of individual memory.[75]

Halbwachs' collective memory does not only reflect the mental space of a social group. Memory is also associated with the material space in which it is produced. In *The Collective Memory*, he explains that in the social framework of a family, the objects of a house recall the family's members. When individuals live on their own, the objects found in their houses recall the very fact that there are no family members to associate with and thus reflect what distinguishes them from others. In this way material objects are themselves unique elements of a society, as they circulate within social groups recalling notions of the past.[76] The form of material objects is exceptionally important, as this is what conveys meaning to the present. Objects are motionless only in the way they appear, and this appearance of stillness lasts for long periods, which is what gives the feeling of stability and consistence to the social group. As material space is intimately associated with the social group that inhabits it, a spatial investigation can reveal aspects and give clues about the structure of the society as a whole. In regard to his urban views, Halbwachs proposes that in small cities, where tradition is strong and the social structure stable, while social groups evolve in time, the material composition of the city changes very slowly. However, if the material composition of the city happens to change drastically, this will have an unavoidable impact on the social group.[77]

The idea that collective memory fades as social groups gradually disappear has been criticised by Jan Assmann. Assmann recognises two types of memory. One is *communicative memory*, which deals with everyday communication and is closely linked to what Halbwachs calls collective memory. It relates to social frameworks and has an expiry date of 80–100 years (three to four generations). The second type is called *cultural memory* and is 'characterised by its distance from the everyday .... Cultural memory has fixed points; its horizon does not change with the passing of time. These fixed points are fateful events of the past, whose memory is maintained through cultural formation (texts, rites, monuments) and institutional communication (recitation, practice, observance).'[78] Assmann termed this new type of memory *cultural memory*, to overcome the paradox arising in Halbwachs' theory, according to which collective memory from the distant past can still be alive even after the disappearance of the associated social group.

Both Halbwachs' theory of collective memory and Assmann's theory of cultural memory involve archival functions in the sense that processes of inscribing, storing and retrieving still take place. As discussed previously, individual memory functions like an internal archiving process. Memories are externalised into a repository realm (wax tablet, memory places, unconscious etc.) to be used later, upon recollection. This function, however, is just as external as the function of collective and cultural memory, as both demand objects and practices to refer to. And so the distinction between individual and collective memory seems to disappear, which supports Connerton's earlier association between Halbwachs' collective memory and Bergson's habit memory.

Notes

1. Jacques Derrida, *Dissemination*, translated by B. Johnson (London: Continuum, 2004), 71–2.
2. Paul Ricoeur, *Memory, History, Forgetting* (Chicago: University of Chicago Press, 2004), 141.
3. *Toute la mémoire du monde*, directed by Alain Resnais (France: Film de la Pléiade, 1957), 35mm.
4. Mike Featherstone, 'Archive', *Theory, Culture and Society*, 23 (2006), 594.
5. In the *Documenta 14* exhibition catalogue Harald Stadler, the director of the Institute of Archaeologies at University of Innsbruck, explains that the roof is not a conventional site for archaeological excavations and proposes a re-evaluation of its potential for research. Being very different from a waste field, the attic is a storage space in which the inhabitants dispose and often forget a range of objects. These objects, when found, can give valuable clues in archaeological works. See Lois Weinberger, *Debris Field; Explorations into the Decrepit 2010–2016* (Kassel: documenta and Museum Fridericianum, 2017).
6. Kabakov's installation was set up for the first time in London between 1985 and 1988. Now it is part of the permanent collection in Nasjonalmuseet for Kunst in Oslo, Norway.
7. Sven Spieker explains that the '"stringing up" of objects was one of the most ancient forms of filing, and the English word "file", which is derived from the French fil (string), originally meant

"to line something up on a piece of string"'. Sven Spieker, *The Big Archive: Art from Bureaucracy* (Cambridge, MA: MIT Press, 2008), ix.

8. Ilya Kabakov, 'The Man who Never Threw Anything Away' in *The Archive,* ed. Charles Merewether (London: Whitechapel Gallery/Cambridge, MA: MIT Press, 2006), 33.
9. Kabakov, 'The Man who Never Threw Anything Away', 35.
10. Hilary Jenkinson, *A Manual of Archive Administration* (London: Percy Lund, Humphries & Co, 1965), 4.
11. Jennie Hill, ed., *The Future of Archives and Recordkeeping: A Reader* (London: Facet, 2010), 7.
12. Jenkinson, *A Manual of Archive Administration,* 8–9.
13. Jacques Derrida, *The Post Card: From Socrates to Freud and Beyond* (Chicago: University of Chicago Press, 1987), 4.
14. Derrida, *The Post Card*, 23.
15. Jacques Derrida, *Archive Fever*, translated by Eric Prenowitz (Chicago: University of Chicago Press, 1995), 91.
16. Herman Rapaport, 'Archive Trauma', *Diacritics* 28: 4 (Winter 1998), 68–9.
17. Derrida, *Archive Fever*, 2.
18. Derrida, *Archive Fever*, 3.
19. Sigmund Freud, *Beyond the Pleasure Principle*, ed. Dufrense Todd, translated by Gregory C. Richter (Peterborough, ON: Broadview Editions, 2011), 129.
20. Derrida, *Archive Fever*, 19.
21. Derrida, *Archive Fever*, 19.
22. In his writings, Freud makes use of the term 'archive' only once. The most common association that he makes is between psychoanalysis and archaeology. In *Constructions in Analysis* (1937), Freud writes: '[the psychoanalyst's] work of construction, or, if it is preferred, of reconstruction, resembles to a great extent an archaeologist's excavation of some dwelling-place that has been destroyed and buried or of some ancient edifice. The two processes are in fact identical, except that the analyst works under better conditions and has more material at his command to assist him, since what he is dealing with is not something destroyed but something that is still alive – and perhaps for another reason as well. But just as the archaeologist builds up the walls of the building from the foundations that have remained standing, determines the number and position of the columns from depressions in the floor and reconstructs the mural decorations and paintings from the remains found in the debris, so does the analyst proceed when he draws his inferences from the fragments of memories, from the associations and from the behaviour of the subject of the analysis.' (Sigmund Freud, 'Constructions in Analysis', *The International Journal of Psycho-Analysis*, 19 (1938), 379.) In the same paper, however, he points out the fundamental difference between archaeology and psychoanalysis. If an archaeological object is destroyed, there is no way that we can get hold of it. On the other hand, when it comes to the psyche of the person '[a]ll of the essentials are preserved; even things that seem completely forgotten are present somehow and somewhere, and have merely been buried and made inaccessible to the subject. Indeed, it may, as we know it, be doubted whether any physical structure can really be the victim of total destruction.' (Sigmund Freud, 'Constructions in Analysis', 380).
23. Derrida, *Archive Fever*, 7.
24. Schwartz and Cook point out that archives have been associated with power since the *mnemons* of ancient Greece. See Joan M. Schwartz and Terry Cook, 'Archives, Records, and Power: The Making of Modern Memory', *Archival Science*, 2 (2002), 3. The *mnemon* was a person trained to memorise and keep track of the proceedings in law courts, as writing had still to be invented. In mythology, the *mnemon* represented a servant of heroes whose mission was to remind them of the divine orders. For example, Achilles was constantly accompanied by a *mnemon* who was assigned to make sure that the former would never kill a son of Apollo, otherwise he would lose his life. However, the myth unfolds in such a way that the reminder fails to function and the *mnemon* himself is put to death. In this sense '[m]emorisation in the Archaic period was ... more than a mere device for keeping facts straight – more than an efficient storage and retrieval system. It was a way of getting (and staying) in touch with a past that would otherwise be consigned to oblivion; it was a fateful fending off of forgetfulness.' Edward S. Casey, *Remembering: A Phenomenological Study* (Bloomington: Indiana University Press, 2000), 12.
25. Derrida, *Archive Fever*, 10–11.
26. In *The Archaeology of Knowledge*, Foucault makes the first attempt in archival theory to dissociate the archive from its spatial dimensions and the common definition as a static repository of memories. Foucault's theoretical approach to the archive is abstract and deals with

'the first law of what can be said, the system that governs the appearance of statements as unique events'. Michel Foucault, *The Archaeology of Knowledge* (London: Tavistock Publications, 1972), 129. The archive reflects the *system of discursivity* and thus decides on the limits of what can be said. Therefore the archive does not reproduce but rather produces meaning and so it can be perceived as an instrument of power.

27. Derrida, *Archive Fever*, 4, note 1.
28. Derrida, *Archive Fever*, 91.
29. In post-Freudian psychoanalysis, the concept of the death drive has received major criticism. According to the theorist Duncan Barford, there are three categories of critics of the death drive, who perceive it as simply redundant, immaterial or morally objectionable. Duncan Barford, 'In Defence of Death' in *The Death Drive: New Life For a Dead Subject?*, ed. Rob Weatherill (London: Rebus Press, 1999), 12.
30. Gilles Deleuze, 'Sadism, Masochism, and the Death Instinct', repr. as an appendix in Freud, *Beyond the Pleasure Principle*, 238.
31. Freud's initial understanding of neurosis was inspired by his studies of biology. He perceived neurosis as a conflict between ego-instincts (self-preservation) and object-instincts (preservation of the species), when the former overrides the latter. As a result, 'the neurotic was engaged in a futile effort to repress the sexual demands of nature'. Barford Duncan, 'In Defence of Death', 17.
32. Freud, *Beyond the Pleasure Principle*, 56.
33. Freud, *Beyond the Pleasure Principle*, 72.
34. Freud used the concept of repetition compulsion for the first time in 1914, in an article entitled 'Remembering, Repeating and Working-Through', and referred to the acting out of some memory that the patient had forgotten.
35. Duncan, 'In Defence of Death', 15.
36. Freud, *Beyond the Pleasure Principle*, 76.
37. Freud, *Beyond the Pleasure Principle*, 77.
38. Freud, *Beyond the Pleasure Principle*, 77.
39. Freud, *Beyond the Pleasure Principle*, 76.
40. Quoted in an appendix to Freud, *Beyond the Pleasure Principle*, 132.
41. Sigmund Freud, *The Economic Problem of Masochism, Part XI (Addenda A, B, and C)*, Standard Edition (London: Hogarth Press, 1924), 164.
42. Sigmund Freud, *Complete Psychological Works*, vol. 22 (1932–1936), *New Introductory Lectures on Psycho-Analysis and Other Works*, gen. ed. James Strachey (London: Vintage, Hogarth Press and the Institute of Psychoanalysis, 2001), 211.
43. Patrick Joyce, 'The Politics of the Liberal Archive', *History of the Human Sciences* 12: 2 (1999), 38.
44. Joyce, 'The Politics of the Liberal Archive', 39.
45. Richard Hornsey, *The Spiv and the Architect: Unruly Life in Postwar London* (Minneapolis: University of Minnesota Press, 2010), 169.
46. Joyce, 'The Politics of the Liberal Archive', 40–42.
47. Joyce, 'The Politics of the Liberal Archive', 44.
48. Joyce, 'The Politics of the Liberal Archive', 47.
49. Mike Featherstone, 'Archiving Cultures', *British Journal of Sociology* 51: 1 (March 2000), 170.
50. Featherstone explains that the flâneur 'is not just the stroller in the city ... Flânerie is a method for reading texts, for reading the traces of the city. It is also a method of writing, of producing and constructing texts.' Mike Featherstone, 'The Flâneur, the City and Virtual Public Life', *Urban Studies* 35: 5–6 (May 1998), 910.
51. This city portrait depicts the unfolding of vivid memories and experiences that both authors obtained in the city and is presented to us in the form of a very rich, enjoyable and highly descriptive travel reportage. It consists of a series of snapshots or, as Benjamin calls them, flashes of light [*Aufblitzen*] from this trip. In 'On the Concept of History', he gives an example of the image of the *flash*. 'The true picture of the past flits by. The past is held back fast only as an image that flashes up at the moment of recognition, never to be seen again. David S. Ferris, ed., *Walter Benjamin: Theoretical Questions* (Stanford, CA: Stanford University Press, 1996), 6. The flash is not an image of the past, but rather the moment that refers to the cognition of that image. For Benjamin, images have a *historical index*. This historical index refers not merely to a certain point in time to which an image belongs, but also the point in time when the image enters into legibility, when it first becomes readable. This moment of recognition is what Benjamin calls the *now of recognisability*. This event establishes a particular relationship

between the past and the present. History cannot belong to the past but belongs, rather, to the present. It becomes legible and readable through a relation of 'what was' with 'now'.

52. Beatriz Colomina, *Privacy and Publicity: Modern Architecture as Mass Media* (Cambridge, MA: MIT Press, 1996), 9.
53. Paul Ricoeur, *Time and Narrative*, vol. 3, translated by K. Blamey and D. Pellauer (Chicago: University of Chicago Press, 1988), 118.
54. Theodor W. Adorno, *Prisms* (Cambridge, MA: MIT Press, 1967), 177.
55. Adorno, *Prisms*, 175.
56. Jussi Parikka, *Digital Memory and the Archive* (Minneapolis: University of Minnesota Press, 2013), 2.
57. Featherstone, 'Archive', 595.
58. Adrian Mackenzie, 'The Mortality of the Virtual: Real-Time, Archive and Dead-Time in Information Networks', *Convergence*, 3: 2 (June 1997), 66.
59. Parikka, *Digital Memory and the Archive*, 84.
60. Although digital recording technologies have developed immensely, their long-lasting archival qualities are not yet confirmed. In fact, Jeff Williams explains that digital archives have 'already shown significant levels of decay', in Nicholas Cullinan, ed., *Tacita Dean: Film* (London: Tate Publishing, 2011), 1.
61. Pierre Nora, *Realms of Memory: Rethinking the French Past*, translated by A. Goldhammer (New York: Columbia University Press, 1996), 8.
62. Timothy D. J. Chappell, *Reading Plato's Theaetetus* (Indianapolis: Hackett, 2005), 172.
63. Frances A. Yates, *The Art of Memory* (London: Pimlico, 2003), 51.
64. Aristotle, *On the Soul, and Other Treatises of Aristotle* (*The Works of Aristotle*, vol. 6), translated from the Greek by Thomas Taylor (Frome: Prometheus Trust, 2003), 81–3.
65. Aristotle, *On the Soul*, 171–8.
66. Yates, *The Art of Memory*, 22–4.
67. John Locke, *An Essay Concerning Human Understanding* (Oxford/New York: Oxford University Press, 2008 [1689]), xii.
68. Anne Whitehead, *Memory* (London: Routledge, 2009), 94.
69. Henri Bergson, *Duration and Simultaneity: Bergson and the Einsteinian Universe* (Manchester: Clinamen Press, 1999), v.
70. Henri Bergson, *Matter and Memory*, translated by M. N. Paul and S. Palmer (New York: Dover, 2004), 92.
71. Whitehead, *Memory*, 127.
72. Maurice Halbwachs, *On Collective Memory*, edited and translated by Lewis A. Coser (Chicago: University of Chicago Press, 1992), 47.
73. Halbwachs, *On Collective Memory*, 47.
74. Paul Connerton, *How Societies Remember* (Cambridge and New York: Cambridge University Press, 1989), 72.
75. Whitehead, *Memory*, 132–3.
76. Maurice Halbwachs, *The Collective Memory*, translated by Francis J. Ditter, Jr and Vida Yazdi Ditter (New York: Harper & Row, 1980), 129.
77. Halbwachs, *The Collective Memory*, 3.
78. Jan Assmann, 'Collective Memory and Cultural Identity', translated by J. Czaplicka, *New German Critique*, 65 (1995), 129.

3
On Fire

> Above all we must recognise the fact that new experience says no to old experience, otherwise we are quite evidently not up against a new experience at all.
>
> Gaston Bachelard, *The Philosophy of No* (1940)

The methodology of fire: Bachelard's model of knowledge production

A study on memory and fire points towards the employment of specific philosophical theories. This is primarily the work of Gaston Bachelard, who is well known for his writings on memory and fragmented time as well as his psychoanalytic approach to fire. Fire's conflicting temporalities of 'life and death' and 'eternal return'[1] lead also to the necessary deployment of a philosophical rival to Bachelard, Henri Bergson. This rivalry is largely because Bachelard strongly opposes the temporal continuity in Bergsonism.[2] On a different level, the theoretical conflict between Bachelard's fragmented time and Bergson's temporal continuity is not accidental, but rather reflects Bachelard's model of knowledge production, which is adopted in this book as a methodology.

In his 1938 book *The New Scientific Spirit*, Bachelard develops his theory of the formation of objective knowledge that is achieved through modern sciences. He disregards natural sciences and alchemy as precursors of modern sciences, claiming that they are products of the imagination and therefore irrational creations of the mind.[3] These creations function like obstacles to the establishment of objective knowledge and therefore the mind has to overcome them. This ongoing effort to overcome them is what generates scientific knowledge itself. This may not be an easy task, as it often requires an 'aggressive' attitude that could potentially shake the foundations of a whole discipline. In fact, a solidly established discipline may require a radical restructuring and a break with its own past.[4] This is what Bachelard calls an *epistemological rupture*. Commenting on the nature of new knowledge, in *The Dialectics of Duration* he writes: 'All knowledge taken at the moment of its constitution is polemical knowledge; it must first destroy in order to make room for its constructions. Destruction is often total and construction never completed.'[5] His views on the formation of knowledge through an 'epistemological rupture' were formed during his work as a secondary school teacher and later on as a university professor. From the early years of his teaching career, Bachelard opposed the French education system. He argued that 'the humanistic *culture générale* of French pedagogy did not stimulate pupils to research: it gave the impression that knowledge was already established and that past texts were the most authoritative'.[6] His answer to this problem was the formation of a *new scientific spirit* that could be attained through the study of positive sciences, such as physics and chemistry, for these sciences were practised with polemic dynamism that opposed past knowledge. Bachelard explains that

> the scientific spirit is essentially a way of rectifying knowledge, a way of broadening the horizon of what is known. Sitting in judgement, it condemns its historic past. Its structure is its awareness of its historical errors. For science, truth is nothing other than a historical corrective for common and primary illusions. The intellectual life of science depends dialectically on this differential of knowledge at the frontier of the unknown.[7]

He traced the emergence of the new scientific spirit to 1905, the year when Einstein published his theory of relativity, and compared it with the established scientific spirit, which was that of knowledge acquired through textbooks. The latter, for Bachelard, ignored the individual passions, interests and creativity of researchers, and stagnated any efforts of collaboration between them. He thus recognised the new scientific spirit as a product of a *dialogued philosophy* that was very different from the monologue that took place within the closed world of philosophy.[8] Cristina Chimisso explains that Bachelard's comparison between the dialogue of science and the monologue of philosophy exposes two of his main philosophical theses:

> First, science is the product of the dialectic between theory and practice: concepts need to become technical to be precise and to develop, and technical applications are performed according to theoretical procedures and on scientific objects, that is, objects created by science. Second, the dialogue between theory and practice is a dialogue between human beings: science is necessarily social.[9]

Scientific knowledge advances by negating the work of the imagination, which is for Bachelard an epistemological obstacle. As the imagination negates rational knowledge, it negates science too. Therefore scientific knowledge is produced dialectically through a ceaseless *no* to its own past, that is a negation to negation, a dialectic reminiscent of Hegel.[10] As such, the dialectics in Bachelard's philosophy are understood in two different ways, both as philosophical dialectic and as dialogue.

In his book *Le Rationalisme Appliqué* (1949), which has yet to be translated into English, Bachelard applies his scientific dialectic for the advancement of knowledge to an ideal school environment. Through dialogue in teaching, the untaught pupil, whose mind is full of imagination, receives a certain degree of knowledge. At the same time the teacher's knowledge becomes more reflective than before. This practice initiates a philosophical dialectic, as knowledge clashes with its negation, which is either the pupil's primary imaginative state of mind or the teacher's previous experience.[11] This understanding of dialectic is different from the Socratic dialectic, the method of maieutics. The latter aims at accessing knowledge that already exists within the person, whereas the former perceives knowledge as a product of the human mind, seeing it therefore as being historically produced. In this example the pedagogical obstacle functions like the scientific one, as it is employed to produce knowledge by negating the creations of the imagination, which are – according to Bachelard – irrationalities of the mind. He argues that 'one can study only what one has first dreamt about',[12] which suggests that attaining the new scientific spirit is the matter of a dream. The first things the human mind dreamt about were natural objects, like fire, water, earth and air, and these are clearly manifest in classical cosmology, natural philosophy, poetry and literature. In *The Psychoanalysis of Fire* Bachelard claims that the 'fire confined to the fireplace was no doubt for man the first object of reverie, the symbol of repose, the invitation to repose.'[13] Hence the starting point of science, which is in essence *real* knowledge, is fire.

In this light, the methodology of this book is based on Bachelard's model of knowledge production. This methodology aims to generate a *conflict*, which is in this case achieved through the dialectic between two opposing philosophical theories. Following Bachelard, however, this conflict carries within itself the memory of fire, as fire is to be found in the very essence of knowledge.

Vernant's interpretation of the Promethean myth

The French historian and anthropologist Jean-Pierre Vernant in *At Man's Table: Hesiod's Foundation Myth of Sacrifice* confirms Bachelard's thesis that conflict carries resonances of fire, and reveals the implicit link between fire and knowledge. His approach is based on the analysis of the myth of Prometheus as delivered by Hesiod (eighth century BC) in *Theogony* and *Works and Days*. According to this myth, when the Olympian gods received a sacrificial meal as a gesture to reconcile themselves with the mortals, the titan Prometheus played a trick against the god Zeus. Zeus had to choose between two of Prometheus' offerings: one that was ox meat hidden inside the animal's stomach, thus having a displeasing exterior, and the other containing bare bones covered in fat and aromatics, therefore having a pleasing exterior. The god chose the latter and from then on mortals received meat whereas the gods received the bones discarded from every sacrifice. This trick enraged Zeus who, in retaliation, hid fire from the mortals. In return, Prometheus climbed Mount Olympus, lit a torch from the sun and brought fire back to them. As a consequence Zeus sent to the humans a counterpart to fire, a '"trap" … a gift no less ambiguous'.[14] This was the first woman, a creation made out of clay by Hephaestus, who was called Pandora.

Vernant suggests that this myth must be read as a single, unified story as opposed to an aggregate of isolated episodes. If one does so, the Promethean fire, which is the first gift offered to humans, can be perceived as a memory of the initial conflict between the titans and the gods. Then, Vernant adds, Zeus' offering of Pandora, the second gift in the myth, is made to counterbalance the theft of fire, as it 'can compensate for fire and provide the balance because she herself is a kind of fire, which will burn men alive by consuming their strength day by day'.[15] Pandora is therefore offered to intensify the conflict between Prometheus and Zeus and, together with fire, establish the mortal status of the human condition.[16] Vernant explains that:

> The myth connects the ritual of sacrifice to primordial events that have made men what they are, mortal creatures living on earth in the midst of countless ills, eating grain from the fields that they have worked, and accompanied by female spouses.[17]

Fire and its counter-manifestation, woman, carry the memory of the divine conflict, which sets the boundaries of human nature. Commenting further on this myth, Vernant explains that Zeus' initial action of hiding fire from humans aimed to deprive them of cooked food. Cooking meat, he points out, distinguishes humans from animals, as the latter only consume raw food. Therefore the Promethean fire, the fire returning this human ability to humans, 'represents culture as opposed to wildness'.[18] The possession of this fire, which is different from the fire of the gods in the sense that the former is perishable and needs to be constantly fed, demands a *technics*, 'a technique of transporting, conserving, and lighting the fire, part of the know-how inseparable from human life'.[19] In this way, Vernant creates the link between fire and knowledge, as it is through the former that humans recognised their technical ability.

Heraclitus' cosmology, fire and conflict

The omnipresence of fire in knowledge and conflict is also confirmed by the ancient philosopher Heraclitus of Ephesus. Heraclitus, like Bachelard, is preoccupied with natural elements and he is known to us particularly as the philosopher of fire. Heraclitus' work dates from the period around 500 BC, some years before the Golden Age of Greece, when issues such as the origins of human nature, the world, the divine presence and the social and economic order were explained according to the culture of religious myths. The rise of natural philosophy marked the beginning of a new movement of thought according to which philosophy was considered as both *investigation* and *instigation*. The natural philosophers, as investigators, examined natural phenomena and tried to give a natural explanation for existing order. At the same time, as instigators, their role was to awaken the sleeping mind of people and elevate their spirit to a higher level of consciousness. Our knowledge of Heraclitus' original doctrine is limited. This is because his writings were offered as a gift to Artemis and deposited in the precincts of her Temple in Ephesus, where they were eventually burnt by Herostratus. Nevertheless, his philosophy is referenced extensively by other ancient philosophers, and the latter's writings are our main sources of his work.

Heraclitus' thought was concerned with questions very familiar to us today. His main inquiry revolved around the way the One manifests itself in the Many and the way the Many obtain within themselves the One. This concept of unity, or of the One, appears in the cosmos as an interaction of four fundamental elements: fire, earth, air and water, with the emphasis on fire as a transforming and generative energy. Specifically,

> fire's alternations: first as sea, and of sea half earth and half lightning dispersed as sea, and measured in the same proportion as existed before it became earth (πυρός τροπαί πρώτον θάλασσα, θαλάσσης δε το μεν ήμισυ γη, το δε ήμισυ πρηστήρ...θάλασσα διαχέεται, και μετρέεται εις τον αυτόν λόγον, οκοίος πρόσθεν ην η γενέσθαι γη).[20]

The significance of fire as the common element of the cosmos is revealed in a different fragment stating that:

> this cosmos (the unity of all that is) was not made by immortal or mortal beings, but always was, is and will be an eternal fire, arising and subsiding in measure (Κόσμον τόνδε, τον αυτόν απάντων, ούτε τις θεών ούτε ανθρώπων εποίησεν αλλ'ήν αεί και εστίν και έσται πυρ αείζωον, απτόμενον μέτρα και αποσβεννύμενον μέτρα).[21]

These last words come to reveal the eternal nature of fire, or of the Logos,[22] and we are introduced to another fundamental topic of his philosophy, the one of measure, or *metron* (μέτρον), which is the power that maintains order in the flux of the cosmos, the balance of forces keeping the earth in constant orbit. For Heraclitus, fire is associated with the concept of time and is constantly in movement, it is eternal. However, this movement is not linear but circular, hence its eternal nature. The three other elements of the fourfold – earth, air, water – are constantly present and are more predominant within space, but fire is the element that brings them into visibility by revealing them. Fire's supremacy is distinct, and when it appears, its most divine and polemic manifestation is lightning, as 'lightning directs everything' (τα δε πάντα οιακίζει κεραυνός).[23]

The controversy surrounding Heraclitus' thought is worth mentioning. As the philosopher Kostas Axelos has mentioned, 'Heraclitus himself is the actualisation of the unity of the opposites.'[24] Famously, Heraclitus supported the idea that 'conflict is universal and that strife is right, and that all things happen through strife and necessity' (πόλεμος πάντων μέν πατήρ έστι, πάντων δέ βασιλεύς. Ειδέναι δέ χρή τον πόλεμον εόντα ξυνόν, καί δίκη έριν, καί γινόμενα πάντα κατ' έριν καί χρεώι).[25] For him, the universality of conflict and strife is central to human nature, cosmic motion and the workings of nature. The opposites, their opposition and their unity, are manifested in a constant motion, as in the case of a river:

> New and different waters flow around those who step into the same river. It disperses and comes together … flows in and out … towards us and away … all things are in process and nothing stays still, and we cannot step twice in the same river (Ποταμοίσι τοίσιν αυτοίσιν εμβαίνουσιν, έτερα και έτερα ύδατα επιρρεί…σκίδνησι καί πάλιν συνάγει…και πρόσεισι και άπεισι. Ποταμοίς τοίς αυτοίς εμβαίνομεν τε καί ουκ εμβαίνομεν, ειμέν τε καί ουκ ειμέν).[26]

The image of the river is a predominant feature in Heraclitus' work. We can see it but we cannot dominate it, as it comes towards us and goes away.

In Heraclitus' ancient philosophy, the cosmos emerged and evolved through the divine interaction of the elements of the fourfold – fire, earth, water, air – with fire as the guiding force. The Heraclitean cosmos is in perpetual flux, for if there is no movement, it ceases to exist. Yet what keeps the world moving is conflict, which is in all cases necessary. The conflict of opposites is the force that leads to the creation of something new: it generates life. This universal and life-giving aspect of conflict is associated with the eternal nature of fire where lightning is its supreme and most polemic manifestation. Accordingly, fire and conflict are interrelated, and are to be found in every dimension, eternally and universally.

What is fire?

In *The Psychoanalysis of Fire*, Bachelard admits that the answer to the question 'what is fire?' is in essence quite vague. In this he concurs with the art historian Alan Krell, who argues that our knowledge of the element has always been ambiguous. He points out that this element, and particularly its manifestation in mythology and religion, is 'largely overlooked'.[27] Any objective definition of fire is blurred by aspects that are coloured by both personal intuition and scientific knowledge.[28] When Bachelard writes, he is struck by the lack of information about fire in scientific books of the day.[29] Whereas once it used to be a fundamental concern for scientific research, nowadays it appears to lack significance, having entered a zone in which its status is taken for granted: our knowledge of fire is already given.

Fire has been an integral part of human life since the early years of mankind, to the extent that both have 'come to resemble one another such that the tread of one tracked the tread of the other'.[30] The moment in history when it becomes part of civilisation marks the arrival of a gradually easier and safer life. At the same time, this arrival implies the beginning of a process where humans start to accommodate such change to their lives. This point in history also marks the beginning of fire's civilising process. From a sociological point of view, in the course of becoming fully human, every individual goes through a civilising process. During this process, people 'learn, largely from others, how to regulate their own sense impressions and impulses, how to behave and how to think. This we may call the civilising process at the individual level.'[31]

The dangers associated with fire are knowledge passed down to us by our elders. We first come to know that we should not touch it. Later, these prohibitions become a result of our direct experience with the element itself. It is in this sense that 'fire is more a *social reality* than a *natural reality*'.[32] Our acquaintance with fire on the level of natural experience occurs, as Bachelard makes clear, in the same way that Prometheus stole fire from Zeus to give it to the mortals on earth. Regardless of the elders' prohibitions, the child will steal fire to experiment, play with, and thus possess knowledge of it. This, according to Bachelard, is the *Prometheus complex*, which 'enables us to understand the interest that is always aroused by the rather trite legend of the father of Fire'.[33]

Fire is most commonly defined in modern encyclopedias as a process of combustion, which manifests itself as heat and light. Compared to earth or water, fire 'is not a substance as such, but a variable combination of substances in a particular and unusual state caused by a chemical reaction'.[34] The impact of fire may be considered catastrophic, as the final products of combustion can never repossess their initial forms and structures. Combustion is by default an irreversible process. The spread of fire has no rules: whatever is flammable will be consumed. Fire causes heat, and heat in turn causes fire. In another mythic representation, this cyclical process is embodied by the phoenix, which is consumed by flames and then reborn from its ashes.

The element of fire carries within it a very peculiar dimension, which does not allow anyone to express a definite opinion or a particular feeling about it. Its impact may be considered catastrophic, as highly organised matter is reduced to a lower state of organisation or integration. Yet its effects are also conducive to processes of reorganisation and reintegration. Destruction by fire provides ground for life to re-emerge. Overall, both good and evil values can be attributed to its presence because:

> It shines in Paradise. It burns in Hell. It is gentleness and torture. It is cookery and it is apocalypse. It is pleasure for the good child sitting prudently by the hearth; yet it punishes any disobedience when the child wishes to play too close to its flames. It is well-being and it is respect. It is a tutelary and a terrible divinity, both good and bad. It can contradict itself; thus it is one the principles of universal explanation.[35]

This ambiguity of emotions and attitudes that accompanies fire is what Bachelard calls the *Empedocles complex,* named after the Greek pre-Socratic philosopher who believed that the four eternal and unalterable elements that constitute the world – fire, air, water and earth – are peacefully united through the opposing, divine powers of Love and Strife.

Fire is self-generating. Through heat generated by combustion, flammable materials within the vicinity will ignite as long as there is enough oxygen. Based on this property, early civilisations made distinctive efforts to proliferate fire as soon as it came under their control. The early men discovered a medium that provided warmth and light, brought them together and also kept predatory animals away, helped them to clear lands and consequently develop agriculture, and also produced more edible and tasty food better suited to human consumption. Thinking of the first people acquainted with fire, Bachelard senses that this element must have been the first object of reverie and a perfect reason for repose. He suggests that a philosophy of repose, or else a philosophy of instants of creation, is inconceivable unless there is a reverie in front of flames.[36] He believes that a child sitting and contemplating in front of a fireplace in our time is an attitude that originates from the past. The child of an earlier civilisation also sat and contemplated in front of flames. This contemplation in both cases is a reverie and not a dream. For Bachelard, the dream develops in a linear way and gradually fades away. The reverie, on the other hand, focuses and reflects on a specific object, it 'works in a star pattern. It returns to its centre to shoot out new beams.'[37]

Fire in mythology

It is often speculated that fire burnt for the first time on Earth roughly 400 million years ago, during the early Devonian period. Before that, the planet did not carry any land plants and was thus incapable of being the site of combustion.[38] Our knowledge of the period in which humans became acquainted with methods of kindling fire is rather unclear. The only evidence that we have of this moment in history are the much later stories narrated by various civilisations. These stories contain no recollection of the actual event but are rather speculations on the origins of human society. Krell claims that the earliest stories of fire relate to the life of Indigenous Australians, which stretch back to roughly 50,000 years ago and are linked with the Dreaming.[39] The Australian curator Wally Caruana explains that the Dreaming is

> the period from the genesis of the universe to a time beyond living memory ... [and] focuses on the activities and epic deeds of the supernatural beings and creator ancestors ... who, in both human

> and non-human form, travelled across the unshaped world, creating everything in it and laying down the laws of social and religious behaviour.[40]

For James George Frazer (1854–1941), a social anthropologist who conducted extensive research on mythology and religion, the stories about fire were myths that should be studied,

> for, while myths never explain the facts which they attempt to elucidate, they incidentally throw light on the mental condition of the men who invented them; and, after all, the mind of man is not less worthy of investigation than the phenomena of nature, from which, indeed, it cannot be ultimately discriminated.[41]

In the *Myths of the Origin of Fire*, Frazer, who undertook the first substantial anthropological research on this topic, collected a broad yet detailed record of myths from all over the world and classified them into three categories. The first refers to the period when mankind was not aware of fire. According to the myths of that age, primitive men suffered from cold weather and their food was eaten uncooked. For instance, the Masingara tribe in New Guinea fed exclusively on bananas and sun-dried fish. Similarly, the Shilluk people in what is now South Sudan also dried their food in the sun. The males of the tribe consumed the upper part of the food – which was somehow cooked – whereas the raw part was given to the females.[42]

The second category refers to the myths of the period when mankind acquired fire but when kindling methods had yet to be invented. These narrations consider lightning to be the initial source of fire, which was sometimes regarded as a divine gift.[43] Myths from this period also refer to the fire created by the branches of trees that are rubbed against each other by the wind. These myths coincide with the description of the origin of fire given by the Roman poet Lucretius.[44] Other myths describe fire being acquired from the sun, the moon and the stars, hence the Greek myth of Prometheus, who stole fire from the sun with a torch to pass it onto the mortals on earth. A few references are made to volcanic eruptions, but there are a vast number of myths that refer to the first fire given to humans as a result of an animal's effort or through the cooperation of many animals.[45]

The third category of myths deals with narrations referring to the period when kindling methods were finally invented. Frazer distinguishes three methods of igniting fire: *the fire-drill, the fire-saw, and the fire-plough* (or *stick-and-groove*). The fire-drill is the method of producing fire when one stick is held upright with its point resting against another that is laid on the ground, and is put in motion through the palms of the hands. The rotating motion of one stick gradually creates a hole in the resting one. After a prolonged effort, the heat created through this friction generates fire in the dry leaves that are placed around the stick.[46] The fire-saw is a similar method of igniting fire where one stick is rubbed against another in a saw-like motion, while the fire-plough refers to the method of rubbing one stick into the groove of another.[47] Considering these three methods of igniting fire, there is a general assumption that could potentially explain the peculiarity or abstraction of certain other myths, which refer to fire being produced by the sixth finger of a woman's hand. Other myths draw parallels between the production of fire through the above-mentioned methods and sexual intercourse between two people. The stick in motion represents the phallus whereas the one at rest represents the female genitalia. This parallelism is embraced up until today and is performed in South Asian ritual practices between the Brahman fire-priest (*Agnihorta*) and his wife to ignite the sacred fire. This is a tradition initiated by the priests of the Vedic religion, an historical predecessor of Hinduism.[48]

In general, myths comprise a body of memories that has survived the passing of time. This body of memories is not the memory of the actual event but rather a reflection of it. A clear illustration of the relationship between myth and memory is found in ancient Greek mythology, specifically in the myth of Memory itself. In Archaic Greece, memory had a mythological dimension with two manifestations. These were *remembering* and *forgetting*, and memory was conceived as both, working hand in hand. It was a dualism of equal parts that united into a single one as a separable pair, the simultaneous working of the mythological deities of Mnemosyne and Lesmosyne. According to Plato, Mnemosyne, the mother of all Muses and the divinity of Memory, is the Goddess who

> first makes [poets] inspired, and then through these inspired ones others share in the enthusiasm, and a chain is formed; for the epic poets, all the good ones, have their excellence not from art, but are inspired, possessed, and thus they utter all these admirable poems. So is it also with the good lyric poets.[49]

Yet Mnemosyne is not merely a source of inspiration. She is also a source of knowledge, 'in principle omniscient',[50] as through her workings poets can access innate knowledge. On the other hand, Lesmosyne or Lethe (forgetfulness) originates from the race of the Night, whose mother is

Discord, and she is the dark element in her family. According to Greek mythology, Lethe is also the name of one of the rivers in the Underworld that brings forgetfulness to the dead.[51] Depending on the version of the myth, dead souls either drink water from the Lethe or bathe in it upon their arrival to the Underworld. Through this process of purification of past memories, souls are prepared to re-emerge in new bodies.[52] In this respect forgetfulness is associated with the image of water. Lesmosyne

> makes everything disappear that belongs to the dark side of human existence. It is only both the elements – giving illumination and letting disappear, Mnemosyne and her counter-pole, Lesmosyne – that make up the entire being of the Goddess, whose name comes solely from the positive side of her field of power. This [is a] union of the opposites under the dominion of the positive.[53]

The image of Memory in Archaic Greece consists of two natures that are diametrically opposite. Their simultaneous pull in opposing directions makes up Memory's complete dimension.

These mythological figures, Mnemosyne and Lesmosyne, are closely related to art, for together they encompass the totality of our cultural legacy. They are a merging of complex histories and cultures, which are presented in the form of an image or figure. In fact, these mythological figures can rightly be considered as works of art because of the effect they have upon language and thought. They are employed as a medium to achieve the disclosure of things, or, as the philosopher Martin Heidegger would describe it, they operate as *techne*. Following Heidegger, '[m]yths let things appear ... they open to occurrences that are otherwise obscure or silent in the absence of words to say them. Myths present to us in "passing light" things that come before us in their own events and that call us to engage perceptively with their events.'[54] This intrinsic quality of myths is for Heidegger equal to poetry. Poetry is projective, it is 'the saying of the world and earth, the saying of the arena of their conflict and thus of the place of all nearness and remoteness of the gods. Poetry is the saying of the unconcealedness of what it is.'[55]

Regarding the image of these mythological figures, Mnemosyne's intention is to *first make the poets inspired*. Poets in turn reflect on Mnemosyne through their songs and chants, and these reflections are her children, the Muses. Accordingly, the divine self-presentation of Mnemosyne seems to lose her image and she comes to be perceived indirectly through this loss. Lesmosyne on the other hand, the river that flows next to the waters of Mnemosyne in the Underworld, refers to human forgetfulness, thus relating to the latter's temporal dimension. In contrast with Mnemosyne, Lethe's image remains visible, as her counter-pole disappears to reveal the entirety of Memory. In this sense, the (imagistic) memory of Mnemosyne suggests that there is always forgetfulness in the process of remembering. In the context of the current discussion, the memory of the event that myths narrate reaches us through the simultaneous workings of Mnemosyne and Lethe. What we receive is both a product inspired by Mnemosyne, which are the myths of the origin of fire, and the outcome of Lethe's counteraction, which is the forgetfulness of the actual event.

Fire in festivities

The memory of fire is not only evident in myths but also manifests itself in rituals and festivities. Historical records indicate that these rituals draw on a tradition of fire festivals that took place in ancient times, long before the spread of Christianity. As described by Frazer in *The Golden Bough,* peasants all across Europe have kindled bonfires and danced around them since time immemorial. Most commonly, these bonfires were lit on certain dates: in the spring (on the first day of Lent, Easter Eve, May Day), midsummer, in the winter on Halloween, Christmas Day, and the Eve of the Twelfth Day (normally referred to as Epiphany). All these celebrations had the character of festivals, as they involved a lot of singing and dancing, and sometimes also drinking and dressing up. The ceremonies most often required the procession of fire around villages and fields, and the burning of wickerwork effigies,[56] which on many occasions were associated with the fertility of the land and cattle, as well as the destruction of witches. Apart from the annual fire festivals, there are also records indicating similar celebrations at irregular intervals. All fire festivals share common characteristics. Their practices are reminiscent of each other and they serve a universal purpose:

> The fire is believed to promote the growth of the crops and the welfare of man and beast, either positively by stimulating them, or negatively by averting the dangers and calamities, which threaten them from such causes as thunder and lightning, conflagration, blight, mildew, vermin, sterility, disease, and not least of all witchcraft.[57]

Frazer offers two explanations as reasons for celebrating these fire festivals. The first supports the argument that these festivals are imitative magic, where the production of fire on earth is intended to recreate the heat and light of the sun in the sky, thus ensuring an endless supply of sunlight on earth. The other theory addresses questions of purification and perceives the production of fire on earth as a means to eliminate demons, witches and monsters, or to cleanse the air. This theory is also shared in the testimonies of people who celebrate the fire festivals. The solar theory has positive attributes, as it is concerned with creativity, whereas the purificatory theory is negative, as it aims at destruction.[58]

Looking at fire's manifestation in myths, rituals and festivities, the confused, ambiguous and uncommon nature of the element becomes apparent. My extensive reference to Frazer's work is not accidental, as he conducted the first and most extensive study on this subject. Bachelard, however, disputes the validity of Frazer's ethnological studies, as the latter's sources are texts as opposed to field observations.[59] In *The Psychoanalysis of Fire,* he openly criticises Frazer, claiming that the explanations he puts forward are utilitarian. Commenting on the Scottish fires of Beltane, for instance, Bachelard writes that Frazer's explanation – where the ashes of the bonfires were used to fertilise the fields – 'introduces a sort of *unconscious rationalisation* which misleads a modern reader who is easily convinced of the usefulness of carbonates and other chemical fertilisers'.[60] Bachelard believes that it is wrong to read primitive practices from the perspective of the modern eye, as these interpretations are in most cases wrong. He is not fully satisfied with a rational and utilitarian interpretation of myths and festivities, and instead he proposes a psychoanalytic approach. Beyond Frazer's practical interpretation, he detects an intimate dream, 'the dream of fertility in its most sexual form. The ashes of the bonfires make fertile both animals and fields, *because* they make women fertile. It is the experience of the flame of love which forms the basis for the objective induction.'[61] He suggests that if a psychoanalytic approach is embraced then many other inconsistencies and curiosities in myths and festivities with regard to fire can easily be justified. Cristina Chimisso understands that Bachelard's attack on Frazer centres on his view that the latter perceived primitive and modern mentalities as homogeneous. This view is opposed to Bachelard's theory of knowledge production through an epistemological rupture according to which the scientific, which is modern, must be discontinuous with the pre-scientific, which is primitive.[62] The practices of primitive societies are guided by desire and imagination, which are the epistemological obstacles that the scientific mind fights to overcome. For Bachelard, any knowledge acquired before the rise of modern science, and this also includes our knowledge of fire, is, effectively, primitive.

Fire in modern science

Bachelard postulates that the epistemological rupture with regard to our primitive knowledge of fire occurred two centuries ago when pre-scientific theories were rejected by radical scientific advancements. Up until that point, there was a general belief, similar to ancient times, that heat was a substance. This prevailed until Antoine Laurent Lavoisier, a natural philosopher from the eighteenth century, conducted for the first time studies in combustion that contradicted the thinking of the era, according to which fire's substance was called *phlogiston*. Phlogiston was considered to be a fundamental ingredient of burnable objects that was released in the air during combustion, with some objects containing more phlogiston than others. The theory of phlogiston was assembled by Georg Stahl towards the end of the seventeenth century as a result of a series of experiments with gases conducted by Robert Hooke, Otto von Guericke and Robert Boyle.[63] The misconceptions regarding phlogiston's validity were mainly shaped by imperfect laboratory conditions. Lavoisier, however, proved that the products of some chemical processes that occur through combustion, rather than being lighter due to the release of phlogiston in the air, are in fact heavier. His theory was further developed by isolating what up until then was called phlogiston and proving that it was simply oxygen. The oxygen involved in the burning process was thought to give weight to the end product. Through his experiments, Lavoisier was the first to draw attention to a new concept that considered heat weightless.[64]

Later on, the physicist Benjamin Thompson, also known as Count Rumford, made a significant step in developing this concept. Through careful experimentation in a cannon factory at the Munich arsenal, he demonstrated that heat is a manifestation of motion. He conducted an experiment by measuring the amount of heat produced through the friction of a dull borer and a cannon barrel that were placed in water. After a period of time, he observed that the water reached boiling point. In a work published in 1797 he states that:

> It's hardly necessary to add that anything which any insulated body, or system of bodies, can continue to furnish *without limitation*, cannot possibly be a *material substance*; and it appears to me to be extremely difficult, if not quite impossible, to form any distinct idea of anything capable of being excited and communicated in the manner in which heat was excited and communicated in these experiments, except it be MOTION [emphasis in the original].[65]

The experiments of Count Rumford and later those of James Joule, an English physicist who developed an accurate theory of the amount of heat produced in relation to the amount of work spent, were the starting point upon which the scientists of the nineteenth century constructed the basis for the theory of thermodynamics. These were initially Sadi Carnot's theory of the macroscopic manifestations of heat in large masses of material and, later, Ludwig Boltzmann, James Clerk Maxwell and Josiah Willard Gibbs' work, which associated the behaviour of large masses with the microscopic movement of atoms and molecules. At the beginning of the twentieth century, Max Planck and Albert Einstein's quantum mechanics filled the gaps with their theories of the radiation of light and heat.[66]

Entropy and thermodynamics

One of the advancements in understanding brought about by the theory of thermodynamics concerns the notion of energy. According to the first law of thermodynamics, which agrees with the pre-existing classical idea, energy is conserved throughout its transformations. The second law of thermodynamics introduces a new concept called *entropy*, which creates a discontinuity with classical mechanics, and refers to the amount of thermal energy unavailable for work within a closed system. Entropy refers to the ability of energy to produce work, and as the latter tends gradually to become minimum, the entropy of a system tends gradually to a maximum. The concept of entropy applied on a cosmic scale implies that the entropy of the universe will tend to a maximum until its heat death. In a 1948 paper, 'Mathematical Theory of Communication', the American mathematician Claude Shannon introduced a new approach to the concept of entropy as he linked it to *missing information*.[67] As soon as the concepts of energy and information were linked together, new directions in the development of information theory, computation and media technology were opened up. What is crucial, however, is that the conceptual merging of energy and information enabled science and technology to become part of the discourse in other disciplines, such as the humanities and the social sciences, as nature and culture were for the first time seen as interrelated. This merging is also crucial in my research, as memory in architecture can be examined in energetic terms.

The archive of fire

Following Bachelard's model of knowledge production, an epistemological rupture occurred during the eighteenth century in studies of combustion, which brought about a radical advancement in our knowledge of the element of fire. Approaching this model through archival theory, one can associate the pre-scientific and scientific knowledge of fire with two separate bodies of knowledge, two distinct archives. As previously mentioned, although archives are repositories of memory, they are also sites of forgetfulness. This is because, according to Derrida, archives manifest violence, 'the violence of power (*Gewalt*) which at once posits and conserves the law'.[68] Therefore whatever is remembered in the archive determines also what must be forgotten. The epistemological rupture in the Bachelardian model of knowledge production occurs when an old archive is replaced by a new one, and when this happens the earlier gradually enters into the realm of oblivion. This justifies why 'our intuitions of fire – more perhaps than of any other phenomenon – are heavily charged with fallacies from the past'.[69] Fire appears to carry with it resonances of pastness that stretch back to ancient times, when it was first invented, but it also emerges as a farrago of scientific material and social inhibitions that are passed on from generation to generation, which together give fire an archival aura.

The association of fire with the notion of the archive allows for a further conceptual leap. Fire can be examined psychoanalytically, for according to Derrida '[t]he theory of psychoanalysis ... becomes a theory of the archive and not only a theory of memory'.[70] This coincides with Bachelard, whose approach to fire is also psychoanalytic. In *The Psychoanalysis of Fire* Bachelard claims that the study of memory cannot be objectively approached as 'the initial charm of the object is so strong that it still has the power to warp the minds of the clearest thinkers and to keep bringing them back to the poetic fold in which dreams replace thought and poems conceal theorems'.[71] This reveals that the difficulty in understanding fire is psychological, which, further bolstered by the fact that this element is no longer a concern for science,[72] justifies Bachelard's move to study it through psychoanalysis. A psychoanalytic approach to the fire archive can therefore confirm certain aspects commonly associated with the element. The concurrent life and death drives with which archives are associated reflect fire's conflicting and obscure qualities of life and death, purity and sin, and creation and destruction respectively.

Notes

1. Gaston Bachelard, *The Psychoanalysis of Fire*, translated by Alan C. M. Ross with preface by Northrop Frye (Boston: Beacon Press, 1968), 7.
2. Gaston Bachelard, *The Dialectic of Duration*, translated by Mary McAllester (Manchester: Clinamen, 2000), 28.
3. Bachelard distinguishes two types of imagination, one that is inspired by novelty and deals with the unexpected and another that explores the depths of being's primitive and eternal aspects. Imagination can find its ultimate expression through poetry.
4. Gaston Bachelard, *The New Scientific Spirit*, translated by Arthur Goldhammer (Boston: Beacon Press, 1985), 173.
5. Bachelard, *The Dialectic of Duration*, 34–5.
6. Cristina Chimisso, *Gaston Bachelard: Critic of Science and the Imagination* (London and New York: Routledge, 2001), 71.
7. Bachelard, *The New Scientific Spirit*, 172.
8. For Bachelard, science that does not engage with theory is pointless, whereas theory that does not engage with practice is metaphysical; thus he insisted on the dialogue of scientific philosophy.
9. Chimisso, *Gaston Bachelard*, 72.
10. The Bachelardian dialectic is reminiscent of Hegel's 'negation to negation' dialectic, yet Bachelard claims that the process that takes place within Hegel's dialectic is closed and determined.
11. Chimisso, *Gaston Bachelard*, 88–9.
12. Bachelard, *The Psychoanalysis of Fire*, 22.
13. Bachelard, *The Psychoanalysis of Fire*, 14.
14. Jean-Pierre Vernant, 'At Man's Table: Hesiod's Foundation Myth of Sacrifice' in *The Cuisine of Sacrifice among the Greeks*, eds, Marcel Detienne and Jean-Pierre Vernant (Chicago: University of Chicago Press, 1989), 23.
15. Vernant, 'At Man's Table', 66.
16. Commenting on Pandora's use in the myth, the contemporary French philosopher Bernard Stiegler explains that this example is very important, as the condition of human mortality is associated with *desire*. This, Stiegler explains, allows one to study mortality psychoanalytically. See Marcel O'Gorman, 'Bernard Stiegler's Pharmacy: A Conversation', *Configurations*, 18: 3 (2010), 459–76.
17. Vernant, 'At Man's Table', 24.
18. Vernant, 'At Man's Table', 38.
19. Vernant, 'At Man's Table', 39.
20. Richard G. Geldard, *Remembering Heraclitus* (Hudson, NY: Lindisfarne Books, 2000), 158, fragment 25.
21. Geldard, *Remembering Heraclitus*, 158, fragment 24.
22. For Heraclitus, the Logos is fire, an amazing source of energy that brings people together for warmth and light and 'it is the one that joins all, the being of all, the being of beings as a whole'. Kenneth Maly and Parvis Emad, *Heidegger on Heraclitus: A New Reading* (Lewiston, NY: E. Mellen Press, 1986), 38.
23. Geldard, *Remembering Heraclitus*, 157, fragment 27.
24. Kostas Axelos, *O Herakleitos kai i Filosofia* (Athens: Exantas, 1976), 38. (Αθηνα: Εξάντας, 1976), 38. (My translation).
25. Geldard, *Remembering Heraclitus*, 158, fragment 18.
26. Geldard, *Remembering Heraclitus*, 158, fragment 21–2.
27. Alan Krell, *Burning Issues: Fire and Art in the Social Imagination* (London: Reaktion, 2011), 51.
28. Bachelard, *The Psychoanalysis of Fire*, 3.
29. Bachelard, *The Psychoanalysis of Fire*, 2.
30. Stephen J. Pyne, *Fire: A Brief History* (London: British Museum Press, 2001), 25.
31. Johan Goudsblom, *Fire and Civilization* (London and New York: Penguin, 1992), 7.
32. Bachelard, *The Psychoanalysis of Fire,* 10.
33. Bachelard, *The Psychoanalysis of Fire,* 11.
34. Philip Ball, *The Elements: A Very Short Introduction* (Oxford and New York: Oxford University Press, 2002), 14.
35. Bachelard, *The Psychoanalysis of Fire,* 7.

36. Bachelard, *The Psychoanalysis of Fire,* 14.
37. Bachelard, *The Psychoanalysis of Fire,* 14.
38. Pyne, *Fire*, 3.
39. Krell, *Burning Issues*, 52.
40. Wally Caruana, *Aboriginal Art* (New York: Thames & Hudson, 1993), 10.
41. James George Frazer, *Myths of the Origin of Fire: An Essay* (New York: Barnes and Noble, 1996 [1930]), 1.
42. Frazer, *Myths of the Origin of Fire*, 202.
43. Frazer, *Myths of the Origin of Fire*, 203–5.
44. Frazer, *Myths of the Origin of Fire*, 205.
45. Frazer, *Myths of the Origin of Fire*, 212.
46. Frazer, *Myths of the Origin of Fire*, 218.
47. Frazer, *Myths of the Origin of Fire*, 222–3.
48. Frazer, *Myths of the Origin of Fire*, 217.
49. Plato, *Ion* 533e quoted in Casey, *Remembering,* 12.
50. Edward Casey, *Remembering: A Phenomenological Study*, second edition (Bloomington, IN: Indiana University Press, 2000), 13.
51. According to the ancient Greek geographer Pausanias, Lethe's spring is located in Boeotia and flows very close to Mnemosyne's spring.
52. Charles E. Scott, *The Time of Memory* (Albany: State University of New York Press, 1999), 33.
53. Karl Kerenyi, 'Mnemosyne–Lesmosyne: On the Springs of "Memory" and "Forgetting"', *Spring: An Annual of Archetypal Psychology and Jungian Thought* (1977), 129–30.
54. Charles E. Scott, *The Time of Memory* (New York: State University of New York Press, 1999), 36.
55. Martin Heidegger, 'The Origin of the Work of Art' in *Poetry, Language, Thought*, translated by Albert Hofstadter (New York: Harper & Row, 1971), 74.
56. The ritual of burning effigies during fire festivals, which derives from an earlier ritual that incorporated human sacrifice, aims to get rid of wicked people through imitative magic. The destruction of the effigy brings about the destruction of the targeted person. In *The Golden Bough,* Frazer discusses the Beltane fires in the Highlands of Scotland. According to him, Highland people were so remote that they were subject to no influence from other cultures, and up until the nineteenth century, they managed to maintain their old heathenism. Julius Caesar described human sacrifices performed by the Celts of Gaul. His observation on Celtic culture agreed also with the report provided by a Greek explorer called Posidonius, according to which, the sacrificial victims of the Celts were criminals or captives from wars. Their sacrifice was a festival occurring every five years, and the number of victims put into fire was analogous to the degree of fertility to be achieved. The ritual was performed by the Druids or priests, and the victims were burnt inside enormous wickerwork images. Apart from humans, the same ritual also included the sacrifice of animals, as they were thought to be disguised witches (familiars). Regardless of who the victims were, the reason for burning them alive was that this was the most efficient method of elimination and purification. On these grounds, the accounts of the Celtic human sacrifices suggest that the fire festivals may have their roots in those years. James George Frazer, *The Golden Bough: A Study in Magic and Religion* (Hertfordshire: Wordsworth Editions, 1993 [1890]), 653–4.
57. Frazer, *The Golden Bough*, 642.
58. Frazer, *The Golden Bough*, 641–3.
59. Chimisso, *Gaston Bachelard*, 167.
60. Bachelard, *The Psychoanalysis of Fire,* 32.
61. Bachelard, *The Psychoanalysis of Fire,* 33.
62. Chimisso, *Gaston Bachelard*, 168–9.
63. Louis-Bernard Guyton de Morveau et al., eds, *A Scientific Correspondence during the Chemical Revolution: Louis Bernard Guyton de Morveau and Richard Kirwan, 1782–1802* (Berkeley, CA: Office for History of Science and Technology, 1994), 24.
64. Guyton de Morveau et al., *A Scientific Correspondence*, 27–8.
65. Quoted in John W. Lyons, *Fire* (New York: Scientific American Library, 1985), 14.
66. Enrico Fermi, *Thermodynamics* (New York: Dover, 1936), ix–x.
67. See Bruce Clarke and Linda Dalrymple Henderson, eds, *From Energy to Information: Representation in Science and Technology, Art, and Literature* (Stanford, CA: Stanford University Press, 2002), 26.

68. Jacques Derrida, *Archive Fever: A Freudian Impression*, translated by Eric Prenowitz (Chicago: University of Chicago Press, 1995), 7.
69. Bachelard, *The Psychoanalysis of Fire*, 3.
70. Derrida, *Archive Fever*, 19.
71. Bachelard, *The Psychoanalysis of Fire*, 2.
72. Bachelard, *The Psychoanalysis of Fire*, 2.

4

On Architectural Conservation

> For, indeed, the greatest glory of a building is not in its stones, nor in its gold. Its glory is in its Age, and in that deep sense of voicefulness, of stern watching, of mysterious sympathy, nay, even of approval or condemnation, which we feel in walls that have long been washed by the passing waves of humanity.
>
> John Ruskin, *The Seven Lamps of Architecture* (1849)

During the last three decades architectural conservation has witnessed exceptional recognition. Whereas once the protection of historic buildings and sites was 'an esoteric interest of a few antiquarians and advocates',[1] today it has developed into an international movement supported by millions regardless of age, profession or social status. The author Eric Russell Chamberlin perceives this recent interest in conservation as

> almost obsessional ... [and points out that] preservation for religious reasons is as old as the first rudimentary tomb. Preservation for aesthetic reasons is as old as civilisation. But 'preservation for preservation's sake', because an object is old, regardless of its religious or aesthetic content, is a thing very much of our day.[2]

This rise in interest can partly be attributed to the emergence of relevant international organisations, such as ICOMOS (International Council on Monuments and Sites) in 1965, and the establishment of UNESCO's World Heritage Convention in 1972.[3]

Although the field of conservation is internationally acknowledged, conservation practice is a phenomenon of the West as it reflects essentially Western values.[4] The Venice Charter, which outlines the internationally accepted policy in regard to the protection of architectural heritage, 'drew heavily on the European background of many of its original authors and focused on the concept of the "monument", a term that encompassed the great historic sites and buildings of antiquity'.[5] The concept of the monument does not reflect transnational values, as some societies do not produce monumental architecture nor necessarily aim to erect permanent structures.[6] It was only after UNESCO's 1994 Nara Document on Authenticity, assembled in Nara, Japan, that the discipline of conservation recognised that non-Western cultures may hold different attitudes with regard to what heritage is and how it is treated. Specifically, Paragraph 7 states: 'All cultures and societies are rooted in the particular forms and means of tangible and intangible expression which constitute their heritage, and these should be respected.'[7] An illustrative example is in Japan, where the ritual dismantlement and reconstruction of Shinto shrines in 20 year periodic intervals contradicts the Western understanding of heritage and permanence[8]; or in sub-Saharan Africa, where cultural heritage is to be found in oral histories transmitted from generation to generation, which – because of their gradual disappearance – are now on UNESCO's conservation agenda.[9]

Focusing on architectural conservation, the discipline received particular attention after the two World Wars. As a common practice following the aftermath of war and to enhance feelings of nationalism, cities and monuments that were severely damaged were often rebuilt from scratch. This reconstruction was sometimes accomplished based exclusively on photographic evidence.[10] A prime example is the reconstruction of the historic centre of Warsaw, most of which was destroyed by Nazi troops in August 1944.[11] Following World War II and taking into consideration the great losses to heritage, UNESCO established the Hague Convention for the Protection of Cultural Property in the Event of Armed Conflict in 1954, which came to strengthen the significance attributed to architectural heritage. After the World Wars and as a result of rapid industrialisation, many historical European city centres in considerably rundown condition went through radical transformation. Transportation and commercial developments combined with extensive urban regeneration schemes paid little respect to the conservation of these centres. In fact, many districts were demolished and rebuilt entirely out of scale, failing to match the surrounding context.

The modern movement, in an attempt to establish a new architectural language freed from the past, disregarded anything that dealt with historic buildings and their conservation. Consequently, modernist architecture brought about a radical distinction between new and old built structures, which further led to a clear distinction between two disciplines: one that dealt exclusively with newness, namely architecture, and one that dealt with oldness, namely conservation. The scope of the first discipline was limited to the exploration of space, whereas the second dealt with the investigation of time. An apparently blind disrespect for heritage caused public reaction, which in the 1970s was expressed with demonstrations and the emergence of campaign organisations such as Save Britain's Heritage and Save Europe's Heritage.[12] More recently, the widening of the cultural heritage remit, which now allows the conservation of entire towns and villages, industrial districts, and twentieth-century architecture including the architecture of the modern movement, has unavoidably contributed to a new wave of interest in heritage.

The designation process

The historical value of buildings or areas is typically acknowledged through designation. The language of designation according to Jorge Otero-Pailos is 'utilitarian: it names buildings and sites as historic resources, as stocks or reserves to be drawn upon when necessary for the purposes of making history … Designation simply names and commits an object to a history yet to be written.'[13] In this sense listed buildings can be used as historical resources for the making of history, and they therefore operate like archives. Part of the process of designation is the issue of a report, which is now a lengthy document that brings together vast amounts of information with regard to the architectural, historical and cultural significance of a building or area. This information is collected in a number of ways and through the employment of innovative new technologies. For instance, the London-based company ScanLAB uses point-cloud infrared laser scanning technology to create 3D models of spaces. According to the founders William Trossell and Matthew Shaw, this technology 'creates detailed pointcloud of 3D information about a site to an unparalleled degree of accuracy … It provides information that can help preserve our cultural heritage.'[14] The intention of ScanLAB is to produce a permanent virtual archive of any built environment before and after conservation. Another type of technology that is particularly popular in conservation is geocoding. Merging cartography and database technology, Geographic Information Systems (GIS) trace the precise location of objects and convert them into simple x,y coordinates. Historic England, a public organisation that looks after England's historic environment, uses GIS technology to confirm the accuracy of old maps or to identify the changes that have taken place when they are compared with the precise location of geocoded buildings.

Designation reports aim at the transformation of a building or an area into a shared document 'in the strict sense of a physical resource from which the public may withdraw the facts of its history'.[15] A clear link between listing and archiving can be found in the creation of a designation report, which is a public written document. A key difference between many building designations and archive entries is that '"reading" some parts of [the historic environment] … can only be achieved through the destruction of the primary record'.[16]

The official archivisation of architecture takes place as soon as a building is designated. At that precise moment, it becomes part of a country's cultural heritage. According to the principles of English Heritage,[17] the historic environment 'reflects the knowledge, beliefs and traditions of diverse communities. It gives distinctiveness, meaning and quality to the places in which we live, providing a sense of continuity and a source of identity. It is a social and economic asset and a resource for learning and enjoyment.'[18] It is thus imperative that every generation respects the historic environment they inherit, as this will benefit them and generations to come. Heritage exists to educate and inspire its audience. Equally, whenever decisions are taken that affect the historic environment, they must be inclusive. It is not only experts who should be involved in this process; everybody should play a part. The management of decisions and actions taken to sustain the heritage values of a historical setting is what English Heritage calls *conservation.*[19]

Changes to buildings

Change in architecture, as in life, is an inevitable fact. When the occupants of a building are not satisfied with it, they tend to move. If a building's efficiency is compromised, certain measures are implemented to adjust it to higher standards. Facades, for instance, can change to accommodate a more contemporary look. This is a fairly common practice in commercial buildings to attract more visitors and gain higher rental value. Equally the mechanical services of a building can change, such as the air-conditioning and ventilation systems, to meet contemporary standards and expectations, and furniture can shift or be replaced. The aspects of buildings that do not change in principle, mainly due to financial restrictions, are the structure and site. Yet exceptions do occasionally occur. One such exception was the relocation of a 750-year-old church to a new site in Germany in 2007 (fig. 4.1).

Figure 4.1 The relocation of the 800-year-old Emmaus church from Heuersdorf in Germany to the nearby village of Borna in 2007. The residents of Heuersdorf, a town that had to be abandoned to allow for the expansion of a nearby coal mine, protested for the safe relocation of their church to a new site. This relocation was a 12-kilometre journey that lasted for nine days. © Susanne Ludwig.

Two of the most striking examples are from Egypt: the relocation of the Temples of Ramses II at Abu Simbel and the Sanctuary of Isis at Philae, organised by UNESCO in the early 1960s so that they were saved from the rising waters of the River Nile. The only buildings that are guaranteed not to change are historic-house museums, such as the Charles Dickens Museum and Sir John Soane's Museum in London. They function by freezing change and always point back to a specific moment in their past. Commenting on the frozen, ghost-like presence of historic-house museums, Paul Eggert claims that although societies are not in need of too many of these museums, they are 'important for they potentially give our embodied lives a sense of chronological amplitude, of connectedness with a continuous history of domestic, civic or grander living than our own'.[20] Buildings that function adequately for their inhabitants are the ones that in most cases survive. On any other occasion buildings are usually made redundant and are eventually demolished. Yet this does not apply to listed buildings. Changes to listed buildings can take place only after special permission is granted. Permission is given following relevant conservation legislation, pertaining to specific qualities and aspects of the building and its history.

Towards the establishment of a modern conservation theory

The understanding of objects and structures from the past as *heritage* and the subsequent application of measures to protect them have evolved in parallel with modernity. Basic restoration ideas were known and maintenance works took place in ancient times.[21] John Stubbs proposes that conservation started as an instinct when humans 'began creating objects and structures that could be used continually'.[22] There is evidence of basic restoration ideas from the early years of Christianity, when Roman temples were restored and converted into churches. The church of San Lorenzo in Miranda, Rome (fig. 4.2), for instance, was originally consecrated as a temple to the late Faustina, wife of Emperor Antoninus Pius in 141 AD.

The first tangible evidence of organised conservation dates back to the Northern Song Dynasty in China (960–1127), during which the collection and documentation of ancient bronzes took place.[23] The modern conservation movement, however, was founded upon fundamental concepts and practices that were born and put into practice during the age of the Enlightenment, when there was a sudden wave of nostalgia for the historic past, as Western societies started to adapt rapidly to the demands of industrialisation.[24] The philosopher Alexander Gottlieb Baumgarten (1714–1762) introduced the word *aesthetics* into German philosophy, referring to the relation between art and nature as perceived through the senses. Towards the end of the eighteenth century, Immanuel Kant developed his theory of aesthetics, according to which beauty was no longer a matter of perfection but rather an empirical sensation.[25] Johann Joachim Winkelmann (1717–1768) made a distinction between an original artwork and its replica, thus formulating an important condition of later conservation principles. Later, D'Alembert and Diderot's *Encyclopédie* (1751–77) illustrated the thirst of man to document and organise his knowledge by archiving it. At the same time, libraries started to become more popular and more inclusive and the quality of printing improved.

Figure 4.2 The church of San Lorenzo in Miranda, Rome, was originally consecrated as a temple in 141 AD and as early as the seventh or eighth century it was transformed into a church. © Uli Gamper.

The modern conservation movement was also influenced by early aesthetic theories of the sublime and the picturesque. The concept of the sublime was first introduced to England after the translation of Dionysius Cassius Longinus's treatise in 1652. *On the Sublime* referred to the 'greatness of conception, elevation of diction and emotional intensity'. It was linked with great, wild, awe-inspiring, and stupendous elements in natural scenery.[26] The picturesque was initially traced in depictions of Italian landscapes and was later associated with the emergence of English gardens. Traditionally English gardens had geometric shapes creating room-like enclosures, but with the emergence of the picturesque they started to obtain an irregular shape and their vegetation was allowed to grow naturally and freely.[27] Eventually, qualities of the picturesque and the sublime were ascribed to ancient architectural creations and natural sceneries respectively in order to ensure their preservation.

Viollet-Le-Duc and the restoration movement

The ideas behind the modern conservation movement were born during the period of the French Revolution. Clearly influenced by concepts developed during the years of the Enlightenment, the French state considered art objects and structures from the past as national heritage. They were credited with an educational status and as a result were under the state's protection. Part of Napoleon's political agenda was to promote himself as emperor of a regime of the same tradition as the Roman Empire. Thus the protection of antique monuments was heralded as a testament to the past Roman glory.[28]

The modern conservation movement moved in two directions: the *restoration* movement in France was led by Eugène Emmanuel Viollet-Le-Duc (1814–1879) and the *conservation* movement in Britain by John Ruskin (1819–1900). In the early nineteenth century, there was great enthusiasm for restoration. The passion for restoration in France was also expressed in debates about the limits of restoration work and whether historic buildings that had suffered damage during the French Revolution and Napoleonic Wars should be repaired. This led to the creation of inventories and the commission of new condition surveys, and the role of the historic buildings and monuments in town planning decisions was re-established.[29] Historic churches, which were the main objects of restoration, were compared to ancient Roman monuments, the only difference being that the latter 'were part of a distant civilisation – "a closed chapter" in history – [that] should be preserved in their present state as a document or as a fragment of a document'.[30] The churches on the other hand, because they related to a living Christian tradition in society, should be restored so that they could be fully operational.

The architect and theorist Viollet-Le-Duc, the most famous restoration personality in France whose work includes Notre-Dame de Paris and La Madeleine at Vézelay,[31] explains in the eighth volume of his *Dictionary of French Architecture*, published in 1866, that 'the term restoration and the thing itself are both modern. To restore a building is not to preserve it, to repair it or to rebuild it; it is to reinstate it in a condition of completeness, which may have never existed at any given time.'[32] Although Viollet-Le-Duc had an exceptional knowledge and understanding of architectural history and practice, he developed his core principles of restoration working on medieval churches.[33] The function of the building, as much as the history of its facade, was significant and worth protecting. Occasionally the restoration of a building to its original condition demanded the destruction of later additions, which themselves were historical. Viollet-Le-Duc's principles on restoration were not clear cut but rather depended upon the architect's critical judgement, interpretation, education and talent as a designer. If, for instance, the existing structure of a building could be improved by replacing it with a stronger and more rigid one, or if advanced materials could prolong the life of the building, then a restoration architect should proceed with these interventions and leave the building *in a condition of completeness, which may have never existed at any given time.* This type of restoration work took place in the new sacristy in the south side of Notre-Dame in 1845, where the timber structure of the roof was replaced by a steel one.[34]

French restoration principles were not limited to France but were widely embraced in many countries. Viollet-Le-Duc was internationally famous, becoming an honorary member of the Royal Institute of British Architects (RIBA) in England (1855) and a member of the Academy of Fine Arts in Milan (1858), and was later honoured by numerous other institutions around the globe.[35]

Restoration in Britain

Restoration in Britain was practised mainly between 1840 and 1860. By 1873, 7144 churches had been restored, of which approximately half were medieval.[36] The most important work was undertaken by the Cambridge Camden Society, founded by John Mason Neale and Benjamin Webb in 1839. Like the French restoration movement, their attention was primarily focused on churches, dealing with a wide range of issues, from the way religion was practised to church building and restoration. Accused of restoration popery, the Society reformed under the name the Ecclesiological Society and concentrated instead on the restoration of English churches to their former glory. Their principles were published in *The Ecclesiologist* in 1842. They pronounced:

> We must, whether from existing evidence or from supposition, recover the original scheme of the edifice as conceived by the first builder, or as begun by him and developed by his immediate successors; or, on the other hand must retain the additions or alterations of subsequent ages, repairing them when needing it, or even carrying

> out perhaps more fully the idea which dictated them … . For our own part we decidedly choose the former; always however remembering that it is of great importance to take into account the age and purity of the later work, the occasion for its addition, its adaptation to its users, and its intrinsic advantages of convenience.[37]

The most famous British restoration architect was George Gilbert Scott. He mainly practised church restorations in a very similar way to Viollet-Le-Duc and published his principles in a paper called *A Plea for the Faithful Restoration of our Ancient Churches*.[38] Scott and Viollet-Le-Duc were two of the most influential restoration architects, but in both cases their principles seemed to contradict their work.[39]

John Ruskin's conservation movement

Following the increasing practice of restoration and as a reaction against it, a new approach to the treatment of historic artefacts came about which was in favour of protection and preservation as opposed to restoration. John Ruskin's publication of *The Seven Lamps of Architecture* in 1849 was a direct attack on the destruction of historic buildings caused by restoration practice and set out the core principles of the conservation movement. Through the seven lamps – sacrifice, truth, power, beauty, life, memory and obedience – Ruskin recognised the moment of an artefact's original creation. As this moment is unrepeatable, the practice of restoration is inappropriate and must be forbidden. In Aphorism 31 in the *Lamp of Memory*, he denounces restoration as

> the most total destruction which a building can suffer: a destruction out of which no remnants can be gathered: a destruction accompanied with false description with the things destroyed. Do not let us deceive ourselves in this important matter; it is *impossible*, as impossible as to raise the dead, to restore anything that has ever been great or beautiful in architecture.[40]

Therefore, when it comes to historic buildings Ruskin detests restoration, for it is 'a Lie from beginning to end'.[41] He proposes to look at buildings as they stand at the time of enquiry. By preserving the building's integrity, the personal sacrifice of everybody involved in its creation is respected too. This way its beauty, which is in essence man's understanding of the beauty of nature, is preserved too. Ruskin's bold assertion was that the glory of a building is to be found in its age, and thus through restoration its true age and glory will vanish.[42] He affirms '*[w]e have no right whatever to touch them*. They are not ours. They belong partly to those who built them, and partly to all the generations of mankind who are to follow us' (emphasis in the original).[43] It is therefore imperative that historic buildings be conserved. On the other hand, buildings constructed in our times can be freely and righteously destroyed, as we are both their creators and destroyers.

Ruskin's theories on conservation were digested slowly. At first the protection of historic buildings was practised and funded by individuals. In 1877 Sidney Colvin, the Slade Professor of Fine Arts at Cambridge, further promoted Ruskin's key points to the public in his article 'Restoration and Anti-Restoration'. In the same year and as a result of this exposure, William Morris founded the Society for the Protection of Ancient Buildings.[44] Morris's views on conservation can be summed up in his term *anti-scrape*, according to which buildings are likened to living beings with a limited lifespan and subject to unavoidable decay. If, however, repairs are necessary, the building should be protected with a minimum effort and without any restoration works.[45] Other individual groups that were formed to protect historic buildings are the Ancient Monuments Society, the Georgian Group and the Victorian Society, all of which are still active today. The first law to be enforced to protect historic buildings was passed on 18 August 1882. This was the Ancient Monuments Act and dealt with pre-historic monuments such as Stonehenge.[46] It was only after the end of World War II that listing became the means to protect historically significant buildings. In France, the conservation movement was also becoming popular and was supported by famous personalities, such as Anatole France, Victor Hugo and Marcel Proust, who openly attacked Viollet-Le-Duc's restoration practice.[47] As a result, the first law dealing with the protection of historic buildings in France was passed on 30 March 1887. This reaction to the British conservation movement against the restoration of buildings spread to the rest of Europe, and since then the United Kingdom has been a leading force in architectural conservation theory and practice.[48]

Ruskin's anti-restoration movement gave negative connotations to the term *restoration* in the English language and the overall practice of dealing with historic buildings has since been called *conservation*.[49] A contradiction arises if one considers that in Latin languages the word *restoration* (*restauro* in Italian, *restauration* in French and so on) has an almost equivalent meaning to the word *conservation* in English. Relativist theorists maintain that there are two possible definitions of conservation.

One follows the Ruskin tradition that opposes the practice of restoration. The other deals with a broader activity that includes restoration amongst other practices and aims also at protection.[50] To avoid confusion, in recent years there has been a reassessment of the words restoration and conservation, and of the relationship between them. Some scholars have proposed a new, combined term, restoration-conservation, which was introduced by the Italian historian Alessandra Melucco Vaccaro in 1996.[51] In North America, the term restoration has come to imply the reintegration of lost styles and the term used has been changed to historic preservation.[52] In this book, the word conservation refers to how it has come to be understood and practised in Britain.

Conservation today

The restoration versus conservation dispute of the nineteenth century was followed by two major events that determined the way conservation is practised today: the publication of Alois Riegl's essay 'The Modern Cult of Monuments' in 1903 and almost a century later the establishment of the Charter of Krakow 2000.[53] Both cases were an attempt to unite the opposing practices of restoration and conservation into a single strategy. Riegl's essay explores the differences between monuments, which are split into two categories: the intentional and the unintentional. An intentional monument is 'a human creation, erected for the specific purpose of keeping single human deeds or events (or a combination thereof) alive in the minds of future generations'[54] whereas unintentional is when its meaning is attributed to a modern understanding and setting and not to its creator's initial intention. Commenting on this distinction, Riegl points out that intentional monuments can at some point also become unintentional, and this is when they outlive what they were created for.

Monuments have value. Unintentional monuments can have historical value, age value, use value and intentional commemorative value. Historical value refers to a monument's inherent capacity to point back to a specific time in history and its 'objective … is … to maintain as genuine as possible a document for future art-historical research'.[55] In this sense historical value allows affected parts to be restored so that the monument's function remains intact. The age value of a monument, on the other hand, contradicts the historical one, as it celebrates its decay and degeneration. Riegl explains that age value becomes directly visible and sensed, and argues that its consideration automatically disregards conservation practice.[56] The contradiction between these two values, which in essence reflects the dispute between restoration and conservation, is resolved on the grounds that if the integrity of the monument is threatened, then the age value can be overwritten by the historical, thus allowing interventions to take place.[57] In other words, restoration is sometimes in favour of conservation when the building itself is threatened. If the intention of the monument's creator was its permanence and longevity, which is what Riegl calls intentional commemorative value, then both conservation and restoration processes are allowed for the monument's revival.

Unintentional monuments can have one more use value, which refers to their usability. Riegl points out that as long as the monument's integrity is not affected, any use can be attributed to it. There are exceptions, however, when 'use value may also require the destruction of a monument; for instance, if decay endangers human life'.[58] Use value therefore legitimises design interventions upon heritage.

As for art value, all monuments possess it as long as they respond to the *Kunstwollen* (artistic will) that is characteristic to each era and subject to contemporary discussion and critique. In general, Riegl attempts to reconcile the debate between restoration and conservation by attributing values to monuments, which can in effect determine any interventions.

The Charter of Krakow 2000 is the latest international document that deals with the same debate. Its implementation follows and revises the objectives outlined in the Charter of Athens 1931 and the Charter of Venice 1964. It was the result of the joint participation at the International Conference on Conservation, 'Krakow 2000', of conservation specialists, academics, government officials and institutions such as ICOMOS and ICCROMM (International Centre for the Study of the Conservation and Restoration of Cultural Heritage). The urgency for the creation of the Krakow Charter is stressed in the preamble of the charter and is attributed to the current state of Europe, which 'is characterised by a cultural diversity and thus by the plurality of fundamental values related to the mobile, immobile and intellectual heritage, the different meanings associated with it and consequently also conflicts of interest'.[59]

The point where the practice of restoration meets that of conservation to act as a single approach is in Article 6:

> [t]he purpose of conservation … is to maintain [the historic buildings' and monuments'] authenticity and integrity, including internal spaces, furnishings and decorations according to their original appearance. Such conservation requires an appropriate 'project of restoration' that defines the methods and aims. In many cases, it also requires an appropriate use, compatible with the existing space and significance.[60]

Conservation practice justifies restorative interventions as long as they achieve the *original appearance* of the treated building. Conservation and restoration processes are thus interrelated and interdependent activities. When it comes to conservation on an urban scale, as in historic towns and villages, in a change of policy from the previous charters, freedom is given for design intervention. Both the Athens and the Venice charters perceived the relationship between the old and the new to be incompatible and thus the distinction between the two had to be visible. Referring to this distinction, Manuel Martín Hernández observes that conservation laws in southern European countries such as Spain, Italy and Greece brought pure facadism to historical districts, and as a result these districts look dull and dysfunctional. He likens this tendency to what Cesare Brandi, one of the most influential personalities in modern conservation theory, has defined 'as a crime of "falsification", a process in which cultural norms force interventions to "cover up" with stylistic elements taken from architectural history'.[61] Instead Article 8 of the Krakow Charter perceives the city as an organic entity, which is 'in the process of continuous evolution and change',[62] freeing interventions from referring to stylistic elements from the past. The only requirement is that, upon completion, they must include a connecting element with the city's organic unity, and this is meant in terms of distinctive dimensions and technological, spatial, decorative and chromatic characteristics.[63] Overall the Krakow Charter 2000 and Riegl's essay share a common aspiration, which is the creation of a single method of approach, evaluation and intervention in heritage.

Bachelard's notion of time and memory in the English heritage protection system

Evidently influenced by the Krakow Charter 2000 and Riegl's essay 'The Modern Cult of Monuments', contemporary English conservation practice follows the policies prescribed in the document published in 2008 by Historic England, discussed earlier in this chapter. It also follows the UK government's planning policies for the historic environment and heritage assets outlined in the National Planning Policy Framework (NPPF) published on 24 July 2018. The core intention of Historic England's document is 'to formalise an approach which takes account of a wide range of heritage values … Although developed primarily to guide the activities of English Heritage staff … [they] commend these Principles, Policies and Guidance for adoption and application by all involved with the historic environment and in making decisions about its future.'[64] Historic England assesses the historic value of a building, monument, site, place, area or landscape according to its heritage asset, which reflects 'a degree of significance meriting consideration in planning decisions'.[65] This assessment is reminiscent of Riegl and the way he categorised monuments according to their value. As soon as the heritage asset is established, designation takes place which 'highlights a building, site or area's special interest and value to this and future generations and gives it protection under law or policy'.[66] The majority of designations are for buildings, and considering their special architectural or historic interest they can be classified into three listing grades:

> Grade I buildings are of exceptional interest. Just 2.5% of listed buildings are Grade I;
> Grade II* buildings are particularly important buildings of more than special interest. 5.8% of listed buildings are Grade II*;
> Grade II buildings of special interest warranting every effort to preserve them. Over 90% of all listed buildings are in this grade.[67]

As a general rule, the older a building is the more chance it stands of being listed. In general all surviving buildings erected before 1700 are listed, as well as the majority of those that were built before 1840. If alterations to listed buildings must take place, Listed Building Consent must first be given, and carrying out work without permission is a criminal offence. The listing status of a building determines how easily consent can be given. For Grade II buildings, alterations and part demolitions may be readily permitted, but when it comes to Grade II* and Grade I buildings, decisions for consent are more complicated. There are occasions when entire areas are designated. These are called conservation areas, and local councils designate them for their architectural and historical interest. Accordingly, interventions and alterations in conservation areas can take place only after local authorities grant relevant permission.

Conservation practice can often expose architectural aspects and qualities that were not intentional or made visible by the original architect. Commenting on this, Otero-Pailos explains that conservationists operate as *creative agents* who challenge the notion of the architect-author.[68] Thus architecture comes to be determined by the contribution of many authors, of the creative agents, with the original architect being just one of many. Depending on the listing grade, some conservation actions are allowed to take place. These can be the periodic renewal of key building elements, repairs that do not affect the current state and look of the building, interventions that may increase knowledge of the past, and restoration practices that will clearly give the building more heritage value than it currently holds. Conservation practice can therefore *freeze* the impact of time on a building, can *reverse* time back to a previous order, or *perform both actions at the same time*. It thus manipulates the impact of time, and effectively the way memory is archived in architecture. Taking this *cronocaotic* function on board, the following paragraphs will discuss English conservation practice through Bachelard's notion of time and memory, and suggest that the staged memories of conservation practice are products of human imagination and creativity.

Bachelard is famous for his work on poetics and the philosophy of science. Although his theory on time and memory is developed throughout his writings, the book that most broadly discusses the issue is *The Dialectic of Duration* (1936). This early book is a direct attack on one of the most popular theories of that period, Bergson's theory of duration. Bachelard accepts everything in Bergsonism apart from continuity.[69] His attack is evident even in the title of the book, which implies that duration must be reduced to discontinuous time and be composed of at least two instants to have a dialectic. In *The Dialectic of Duration* Bachelard expands on his philosophy of repose, developed in opposition to Bergson's duration which cannot be broken down to instants, points and intervals. The philosophy of repose punctuates duration, thus developing a discontinuous Bergsonism. When it comes to human evolution, duration is marked by instances of creation resulting from the dialectic between being and non-being, nothingness. In *Creative Evolution*, Bergson gives priority to positive actions by saying that no truly negative actions exist. Therefore when it comes to words, negative ones have significance only in relation to the positive words they negate.[70]

Bachelard, whose understanding of dialectics goes back to Hegel, rejects this imbalance of the dialectic by referring to the example of fullness and emptiness. He explains that for Bergson, the meaning of the word *empty* comes from the verb, the action *to empty*, and is considered as a positive action when the emptiness it brings about is simply the imaginary or actual disappearance of a substance, whereas the disappearance itself, *the emptiness*, has no significance in understanding the action. Bachelard, however, argues that something can be emptied only if it was previously full, and inversely something can be filled up only if it was previously empty. Both actions can thus be understood only in relation to each other and therefore the dialectic of fullness and emptiness is in perfect balance. Commenting on the evolution of philosophy, the duration of which is marked by creative instants of being and non-being,

Bachelard argues that 'pure thought must begin by refusing life. The first clear thought is the thought of nothingness'.[71] Developing out of nothingness, philosophy and knowledge in general unfold in time in alternating periods of affirmation and negation. Accordingly, Bachelard explains, 'there is a fundamental heterogeneity at the very heart of lived, active, creative duration, and that in order to know or use time well, we must activate the rhythm of creation and destruction, of work and repose'.[72] The perception of time is thus a matter of following the *rhythm* of instants created by periods of work and repose.

The concept of rhythm for Bachelard is different from Bergson's. For the latter rhythm in music and in life is used as an analogy to explain duration, whereas for Bachelard rhythm exists so that evolution can take place. Rhythm, for Bachelard, is

> driven by the will to perpetuate or 'copy' ourselves. However, each time we set ourselves to this task, we must exceed or improve on this 'copy', or else it will dissolve itself … . The moment of remaking breaks the flow of time – repetition, to survive, must involve a negation, as well as an affirmation, of previous repetitions. Rhythm is not an 'eternal return', but an 'eternal reprise'.[73]

Based on studies of the phenomenology of rhythms by the Brazilian philosopher Pinheiro dos Santos, Bachelard explains that matter has wave and rhythmic characteristics. For example, the exchanges of energy between different types of chemical matter unfold in time, undulating to the rhythms of specific frequencies. According to kinetic theory, solids are stable due to conflicting rhythms: 'Our houses are built with an anarchy of vibrations. We walk on an anarchy of vibrations. We sit down on an anarchy of vibrations. The pyramids of Egypt, whose function is to contemplate the unchanging centuries, are endless cacophonies.'[74] So, given that solids are composed of vibrations, if they stop vibrating, they cease to exist. Nevertheless, it is not only matter that vibrates but time too. When matter is resting in time, which vibrates, it has energy even in repose.[75] Rhythm is not just present in matter but in life too. In fact, due to its complexity, life is more discontinuous and rhythmic than matter. And amongst all life energies, mental energy, 'given its fragility and complexity, is … the closest to undulating and quantic energy, the most marked by vibration and rhythm'.[76]

Bachelard's understanding of memory is influenced by the work of Maurice Halbwachs and the French psychologist and philosopher Pierre Janet. For Bachelard, memory is discontinuous and defined by consecutive breaks. We remember past events as a series of decisive instants. Therefore narrating our personal history is in essence a narration of our disconnected actions, which make up our entire past. This association of memory with narration follows Janet, who claims that 'What has created humanity is narration, and not by any means recitation.'[77] For Janet, recitation refers to the beginning of human memory, according to which a memorised scene of something existing is recited so that relevant information can be transmitted to all interested parties. Narration, although similar, refers to events that have already finished and disappeared. In any case, both recitation and narration demand the practice of memory. When one finds it difficult to recall a memory, one has to pay additional attention to the instant, the framework, in which that specific memory is fixed. In *The Poetics of Space*, Bachelard writes: 'Memories are motionless, and the more securely they are fixed in space, the sounder they are. To localise a memory in time is merely a matter for the biographer and only corresponds to a sort of external history, for external use, to be communicated to others.'[78] At this point Bachelard draws on Halbwachs' philosophy, according to which memory is a construct dependent on the social framework in which it is created. As discussed in chapter 2, the past for Halbwachs does not recur in the way it happened but is instead reconstructed in the present. As the current social context is different from the context in which memories of the past were made, these memories can never reconstruct the past the way it was. A social framework is therefore fundamentally crucial to the process of recollection. Regarding the preservation of memories, attention has to be paid to the actual fixing of memory. Memories cannot relate to their frameworks if 'there is no spoken, expressed, or dramatised fixing … Reflection must construct time around an event at the very moment when the event takes place so that we can rediscover the event in the memory of time that has disappeared.'[79]

Memory for Bachelard is also affected by deferred behaviour, when actions are postponed for a later moment. His views on deferred actions are influenced by Janet, who claims that postponing an action for a later moment in time is the origin of human memory. Postponing an action reveals memory's social aspect. To illustrate this argument, Bachelard uses Janet's example of sentries in military camps. If a sentry witnesses the invasion of an enemy, he defers the fact, the invasion, by suspending it in his memory until he runs back to the camp to inform the soldiers about it. The deferral of an action implies that its beginning and end are controlled by us; thus its duration is relative. Commenting on deferred actions, Bachelard claims that 'we remember an action much better by linking it to what follows it rather than to what precedes it'.[80] A deferral creates a temporal void before an action takes place. Equally, a void is also created by the expectation of desired events: time feels empty while we wait for these events to take place. Temporal voids create the framework for new memories to be acquired and fixed. This way the deferral of actions and anticipation of events make memory discontinuous, as it is composed of temporal breaks or voids. Conrad Russell argues accordingly that Bachelard's memory, compared to Bergson's photographic record of pure memory, 'is more cinematic, with jumps, cuts and editing'.[81]

Bachelard's memory is also *fictive*,[82] as it is 'creating or created by imagination'.[83] In order to remember, one has to actively pay attention to the frameworks in which memories are fixed. The same happens with narrations. One has to practice with the memory before narrating past events in the correct order. Equally, the discontinuity created by the temporal voids of deferred actions and the anticipation of events demand the work of a creative mind, which, as in cinematographic processes, jumps, cuts and edits so that memory can be assembled. Active creativity is also found in time. The dynamism of Bachelardian time is a result of the rhythm of instants that are created by periods of work and repose, and this is a product of human imagination too. Imagination is a product of the creative faculty of the mind, which is different from the simple reproduction of perception. Whereas traditionally imagination is understood as the faculty that deals with the formation of images, for Bachelard, imagination is the faculty that deforms and eventually liberates us from the initial images of perception. Commenting on Bachelardian imagination, Edward Kaplan explains that this 'freedom from a mental imitation of reality – e.g., sight – [is what] Bachelard calls the "function of the unreal", the imaginative force which enables man to create new images instead of adjusting to reality as given'.[84]

Throughout his philosophical career, Bachelard's views on the imagination changed significantly. Initially, in *The Psychoanalysis of Fire* (1938), he approaches the concept of imagination through Freudian psychoanalysis. In later works, as he finds psychoanalysis a restricting experience, he distinguishes it from psychology[85] and turns towards 'the zone of active imagination, where man is a thinking, willing being, [and] an openness is retained. "Here the depths are not so fearful nor the heights so unattainable." Here we can retrace, reclaim, retrieve, relive, and even transform experience in our imaginative selves.'[86] Like any other process that unfolds in time, imagination occurs as a result of dialectics: 'a dual participation of desire and fear, a participation of good and evil, a peaceful participation of black and white'.[87] In *Water and Dreams* (1942),

Bachelard distinguishes two types of imagination: one is inspired by novelty and deals with the unexpected, and the other explores the depths of the human's primitive and eternal aspects. This is what he calls formal and material imagination respectively.[88] For instance, when it comes to water, one has to look beneath the imagination of its form and meditate upon the imagination of the substance itself. This is when the person experiences what is 'not infinity ... but depth'.[89]

The way to meditate on an object, Bachelard explains, is to be in a state of reverie, which allows for interchanging states between the real and the imaginary. During a reverie, imagination finds its ultimate and unlimited expression though poetry. It becomes a 'phenomenon of freedom'[90] that disconnects us from the past and present and looks at the future. Later, in *The Poetics of Space* (1958), he clearly distances himself from psychoanalysis and psychology, and approaches the study of imagination through phenomenology; hence the image is no longer treated as a reduction of what it really is but as an exaggeration.[91] This method entails a prolonged exaggeration of the image so that one can enter into it. Bachelard examines the image of the oneiric house and other key images, such as drawers, chests, wardrobes, nests and shells, which are all similar in the sense that they 'are interrelated in their common evocation of a relationship of intimacy and refuge'.[92] Lastly, in *The Poetics of Reverie* (1960), he treats imagination from a Jungian perspective. Based on Jung's understanding of the androgynous nature of the human psyche, the *anima* and the *animus*, Bachelard explains that the reverie through which imagination finds its ultimate expression reflects the feminine aspect of the human psyche for both sexes, and is different from the dream, which is masculine. He emphasises the difference in gender of the words *dream* and *reverie* (in French, respectively *le reve* or *le signe*, masculine, and *la reverie*, feminine). In Bachelard's approach to imagination there is thus a radical shift, or in Bachelard's own terms an epistemological rupture, in which his epistemological interests are replaced by metaphysical ones. The same shift occurs in his understanding of the reverie, which comprises a key concept in his philosophy. Reverie, which is the creative daydream, is initially treated with '*malign vigilance* [but then develops into] a joyous and admiring ... poetic reverie'[93] reminiscent of Husserl's phenomenology (italics mine).

To return to Bachelard's notion of memory: its fictive quality, the fact that it is 'creating or created by imagination', can be associated with contemporary English conservation practice. The degree and nature of intervention that can take place in a designated building depend upon its listing grade. Some elements of the building might be subject to periodic renewal. Repairs that do not affect the state or external look of the building, and interventions that increase knowledge of its past, might be permitted to take place. Restoration works that give higher heritage value to the building are potentially welcome. All these actions have a profound impact on time, as they tend to either freeze and/or reverse it, which affects the way memory functions. Conservation and restoration processes punctuate the evolution of a building at specific points in time. The building's evolution is therefore broken down into instants and its duration becomes discontinuous, which conforms to Bachelard's understanding of time.

Following his philosophy further, matter has wave and rhythmic characteristics. The same applies to buildings. Conservation practice comes to capture and preserve a building's vibration at certain points in time, for if this vibration stops, the building will cease to exist. It thus exposes key instants of the building's evolution and this introduces discontinuity, which, according to Bachelard, triggers memory.

Conservation practice aims at fixing memories in space, such as in a building, where these memories can be kept motionless, as 'the more securely they are fixed in space, the sounder they are'.[94] Once memories are securely fixed in space, the spectators need to actively engage by employing their imagination. Memory's fictive aspect is thus revealed, as the work of creative minds is necessary, which, like in cinematographic processes, jumps, cuts and edits so that memory can be accessed and remembering can take place.

The Bergsonian archive and the need to reassess the listing process

During his life Henri Bergson developed a body of work that by the 1920s had become one of the most influential philosophies in Europe. His radical views on time, memory, evolution and difference were embraced widely, to an extent that as John Mullarkey notes 'he was at one point held to be both "the greatest thinker in the world" and "the most dangerous man in the world"'.[95] Although Bergson's main concepts were absorbed into later philosophical movements, such as phenomenology and structuralism, Bergsonism as a philosophy gradually lost its momentum until the 1990s, when there was a sudden revival attributed mainly to the work of Gilles Deleuze. One of the fundamental points in Bergson's philosophy that is also attacked in Bachelard's *The Dialectic of Duration* is the concept of duration or temporal continuity. Whereas for Bachelard time is fragmented, for Bergson there are two types: *homogenous time*, or the time of physics, which is made up of instants and we measure in space, and *real time*, or *duration*, which is heterogeneous and can only be grasped with the method of intuition. In *Time and Free Will*, Bergson's doctoral thesis, he explains that reality can be perceived in two ways. Firstly we can experience a homogenous reality, namely space, which is 'clearly conceived by the human intellect, [and] enables us to use clean-cut distinctions, to count, to abstract, and perhaps also to speak'.[96] Space, which is a homogenous medium, is measurable. Equally, if time is projected in space, it too can be measured. It can have a number. Yet for Bergson this is not real time, but a human construct perceived in space. The real time, duration, takes place in the second kind of reality, a reality that is heterogeneous and comprises a multiplicity of states of consciousness or of sensible qualities. In *Duration and Simultaneity*, Bergson manages to elucidate the difference between duration and measurable time by using an example of someone drawing a line on a piece of paper. If this action is performed with the eyes shut, the motion is perceived from within, as something of his own flow. This is duration. Repeating the same example with the eyes open, one can actually see the line being traced on the piece of paper. This line is perfectly dividable and measurable, and time is measured in space through the intermediary of motion.[97]

Duration is perceived with intuition. Bergson adopted the latter as his philosophical methodology. Referring to intuition, Deleuze in *Bergsonism* writes that intuition 'is neither a feeling, and inspiration, nor a disorderly sympathy, but a fully developed method, one of the most fully developed methods in philosophy. It has its strict rules, constituting that which Bergson calls "precision" in philosophy .... Intuition presupposes *duration*.'[98] Intuition is a method of division and is inspired by the Platonic model.[99] The latter divides a composite into halves, which implies that in the selection process one of the two halves must be isolated. In this sense, not knowing which correct half to choose and having no alternative, the Platonic model of division can be considered insufficient and ineffective. The Bergsonian model by contrast splits composites into *tendencies*. One tendency is duration, which includes all differences in kind, and another tendency is space, which includes all differences in degree. Thus the Bergsonian model makes a qualitative division, overcoming the dilemma of the Platonic model. The division of a composite is not a straightforward process. If one considers the composite of experience, for instance, and divides it into its conditions, one will find that perception and recollection always interpenetrate each other. Thus perception and recollection differ only in degree and not in kind, quantitatively and not qualitatively. To achieve the *pure division* of experience, the composite must be split into its qualitative and quantitative tendencies, which are duration (time) and expansion (space) respectively: the famous Bergsonian dualism.[100] Pure duration implies a continuous unfolding, a process that is purely internal, whereas space implies an external articulation that has no continuity. This is what an experience from the past suggests. It is a recollection of space by a conscious mind that endures.

Bergson's method of intuition seeks experience at its source, traces its pure essence and reveals a difference in kind.[101] Yet intuition is not duration itself, it is 'the movement by which we emerge from our own duration, by which we make use of our duration to affirm and immediately recognise the existence of other durations, above or below us'.[102] It is in this sense that Bergson uses intuition as a methodology: to access philosophical knowledge. This is because through the time of physics, or the Bachelardian time of instants, one can only acquire scientific knowledge, as instants are scientific constructs. The problem of scientific knowledge, Bergson explains, is that although it is a product of the intellect it is insufficient to explore and understand life, and thus philosophy cannot rely on it. He writes that the 'intellect is characterised by a natural inability to comprehend life'.[103] Instead the intellect can comprehend only a segment of reality that is inert matter. In *Creative Evolution* he adds that 'the intellect is never at ease, never entirely at home, except when it is working upon inert matter, more particularly upon solids'.[104] Bergson explains further that the intellect cannot perceive any organic, fluid or animated matter. In the process of understanding movement and fluidity the intellect 'starts from immobility, as if this were the ultimate reality: when it tries to form an idea of movement, it does so by constructing movement out of immobilities put together'.[105] It therefore perceives reality as immobile, and any fluidity or movement escapes it.

The concept of memory is resonant throughout Bergson's work. One of his early books, *Matter and Memory*, deals with it in detail. In this book, which as James Burton points out has not yet been fully appreciated in fields that deal with the topic of memory at their core,[106] Bergson makes a clear distinction between memory and perception, claiming that the former must no longer be considered as a *loose* manifestation of the latter. He writes that:

> If we make recollection merely a weakened perception, we misunderstand the essential difference between the past and the present, we abandon all hope of understanding the phenomena of recognition and more generally the mechanism of the unconscious. But

> inversely, if recollection is regarded as weakened perception, perception must be regarded as a stronger recollection.[107]

The reason that the two are mixed up is that they are the tendencies of a composite that is consciousness. Describing the relationship of these tendencies, Bergson claims that the 'two acts, perception and recollection, always interpenetrate each other, are always exchanging something of their substance as by a process of endosmosis'.[108] He examines the relationship between memory and perception in the context of the living body. When the senses are open, they perceive matter, which is an aggregate of images. Yet amongst all images, there is one image, the image of the body, that we know not only because we can perceive it as matter but also because of our feelings: we know it from within.

The body-image is different from other images, which we can call image-objects, because it deals with movement. It can move or be moved by image-objects. Thus the body is considered the centre of action.[109] The relationship between body, matter and perception is that 'matter [is] the aggregate of images, and perception of matter [is] these same images referred to the eventual action of one particular image, my body'.[110] Therefore, because the perception of matter, or else the image-objects, depends on the body, perception itself is limited. This is because the body-image cannot absorb the totality of image-objects, as this would demand pure perception. Instead it absorbs only a fraction that is of interest and dictated by its past experience, memory.[111] Therefore perception filters out the totality of image-objects based on memory, and this affects consciousness too.

For Bergson, there are two main types of memory: *the habit* and *representational memory*. As discussed in chapter 2, the former deals with a memory that is acquired through the repetition of an action. This action, like learning a text by heart,

> demands first a decomposition and then a recomposition of the whole action ... it is stored up in a mechanism which is set in motion as a whole by an initial impulse, in a closed system of automatic movements which succeed each other in the same order and, together, take the same length of time.[112]

Representational memory is memory that 'records, in the form of memory-images, all the events of our daily life as they occur in time; it neglects no detail; it leaves to each fact, to each gesture, its place and date'.[113] In order to produce a memory-image in the present, one needs to trace it amongst the totality of all memory-images that the memory has absorbed and recorded, and bring it forth into the present. Recollection is therefore the actualisation of a memory-image from a *virtual place* or substratum. In *Bergsonism,* Deleuze illustrates graphically how representational memory takes place, like a detachment from the present with a leap in the past. Through a process of adjustment, like the focusing of a camera, we approach certain regions of the past, and gradually this virtual state of recollection becomes actual. From the general past we leap into a particular past.[114] Accordingly, as soon as our needs and interests dictate the tracking down of a particular memory, the filtering out that takes place in this process of actualisation can in fact be perceived as a process of forgetting.[115]

In the same way that our perception of things depends upon the place where these things are, the past can only be perceived in the place where it exists, which is in the past itself and not in the present. Thus the recollection of a memory-image appears as if it came out of some sort of storage place. Nevertheless as James Burton points out, this is 'a false impression … it is only in the process of recall that memory images acquire this status, being made to look like they have come from an archive.'[116] Commenting on the production of representational memories, John Mullarkey claims that Bergson's memory is therefore not dualistic but in fact tripartite.[117] This third type of memory is called *pure memory* and refers to the totality of memory-images found in a virtual state.

Pure memory is one of the most radical aspects in Bergson's philosophy. It suggests that the entirety of our experience is constantly being recorded in some sort of virtual substratum, and that this record bears no materiality and occupies no physical space. The Bergsonian concept of duration collapses the distinction between past and present and thus memory-images are to be found in a continuous, ongoing lingering until the moment they are actualised. The mental process of recollection occurs with a movement from the virtual to the actual, hence the illusion that memory-images are *archived* somewhere in the brain. Burton remarks that 'when we *recall* something we are writing it on a wax tablet, rather than when we wish to memorise it, since it is only in making the recollection a part of the present that we actualize it'.[118] It is in this sense that the Bergsonian theory of memory is non-archival, for memory is no longer associated with the recording and storing of information. The reversal in this conventional understanding of memory has implications for the nature of archives too. If memory is imprinted *only* in the process of recall, when it becomes actualised, then archives appear *only* when memory is actualised, during the process of recollection. Effectively, the totality of memory does not reside in archives but elsewhere; thus whatever is to be found in archives is merely a reduction of memory. In this sense, *archives are sites of forgetting*.

The reduction of memory associated with the notion of the archive can equally be applied to listing. Whereas listing aims at the protection of a building's memory, if considered in Bergsonian terms, this is no longer valid. Conservation practice takes the form of a body-image that, depending on its needs and interests in the present, filters out or contracts the totality of memory-images that comprise its entire past, thus preserving only an aspect of it. This way the very process of listing, which is in essence an archiving process, is elusive, as the building's memory vanishes. Listed buildings can therefore be considered as sites of forgetting that negatively challenge the main purpose of conservation and raise questions in regard to its theoretical validity and sound practice.

Notes

1. John H. Stubbs, *Time Honored: A Global View of Architectural Conservation: Parameters, Theory and Evolution of an Ethos* (Hoboken, NJ: John Wiley & Sons, 2009), 8.
2. E. R. Chamberlin, *Preserving the Past* (London: Dent, 1979), ix.
3. The significance of UNESCO's World Heritage Convention is 'that it links together in a single document the concepts of nature conservation and the preservation of cultural properties. The Convention recognizes the way in which people interact with nature, and the fundamental need to preserve the balance between the two' (http://whc.unesco.org/en/convention/, accessed 2 August 2018).
4. It is worth mentioning, however, that in the past not everyone in the West shared the same views on permanence. Referring to the buildings of East Berlin, the architectural historian Florian Urban points out that throughout most of the twentieth century the buildings there had a pre-determined lifespan. This was fixed by politicians and urban planners, who proclaimed that buildings do not only deteriorate within a few decades but also become obsolete; therefore they had to be demolished and built anew. Florian Urban, 'From Periodical Obsolescence to Eternal Preservation', *Future Anterior*, 3: 1 (Summer 2006), 26.
5. Graham Brooks, 'The Burra Charter: Australia's Methodology for Conserving Cultural Heritage', *Places*, 8: 1 (1992), 84.
6. Seung-Jin Chung, 'East Asian values in historic conservation', *Journal of Architectural Conservation: The International Journal for Historic Buildings, Monuments and Places*, 11: 1 (March 2005), 59.
7. Nara Document on Authenticity, https://whc.unesco.org/document/116018, accessed 2 August 2018.
8. Jukka Jokilehto, *A History of Architectural Conservation* (Oxford and Boston: Butterworth-Heinemann, 1999), 278.
9. 'Oral Traditions and Expressions including Language as a Vehicle of the Intangible Cultural Heritage', http://www.unesco.org/culture/ich/index.php?pg=00053, accessed 4 July 2018.
10. Aylin Orbasli, *Architectural Conservation: Principles and Practice* (Oxford and Malden, MA: Blackwell Science, 2008), 21.
11. 'Historic Centre of Warsaw', http://whc.unesco.org/en/list/30, accessed 30 April 2018.
12. Orbasli, *Architectural Conservation*, 25.
13. Jorge Otero-Pailos, 'Mnemonic Value and Historic Preservation' in *Spatial Recall: Memory in Architecture and Landscape,* ed. Marc Treib (New York and London: Routledge, 2009), 245–6.

14. Quoted in Merlin Fulcher, 'ScanLAB Video: The AJ Office in 3D', *The Architects Journal*, 233: 24 (2011), 8.
15. Otero-Pailos, 'Mnemonic Value and Historic Preservation', 247.
16. English Heritage, 'Conservation Principles, Policies and Guidance' (April 2008), 54, principle 123, https://www.historicengland.org.uk/images–books/publications/conservation–principles–sustainable–management–historic–environment/, accessed 30 April 2018.
17. Up until April 1, 2015, Historic England was commonly known as English Heritage.
18. English Heritage, 'Conservation Principles, Policies and Guidance', 19, principle 1.2.
19. English Heritage, 'Conservation Principles, Policies and Guidance', 22, principle 4.2.
20. Paul Eggert, *Securing the Past: Conservation in Art, Architecture and Literature* (Cambridge and New York: Cambridge University Press, 2009), 21.
21. Orbasli, *Architectural Conservation*, 16.
22. Stubbs, *Time Honored*, 25.
23. Stubbs, *Time Honored*, 25.
24. Stubbs, *Time Honored*, 57.
25. John Macarthur, *The Picturesque: Architecture, Disgust and other Irregularities* (London: Routledge, 2007), 3–4.
26. Jokilehto, *A History of Architectural Conservation*, 50.
27. Macarthur, *The Picturesque*, 4.
28. Jokilehto, *A History of Architectural Conservation*, 69.
29. John H. Stubbs and Emily Gunzburger Makas, *Architectural Conservation in Europe and the Americas: National Experiences and Practice* (Hoboken, NJ: John Wiley & Sons, 2011), 41.
30. Jokilehto, *A History of Architectural Conservation*, 150.
31. La Madeleine at Vézelay entered the UNESCO heritage list in 1979.
32. Eugène Viollet-Le-Duc, 'On Restoration' in *Dictionnaire Raisonné de l'Architecture Francaise du XIe au XVIe Siécle*, vol. 8, as cited in M. F. Hearn, ed., *The Architectural Theory of Viollet-Le-Duc: Readings and Commentary* (Cambridge, MA: MIT Press, 1992), 269.
33. Stubbs, *Time Honored*, 215–16.
34. Jokilehto, *A History of Architectural Conservation*, 155.
35. Jokilehto, *A History of Architectural Conservation*, 155.
36. Eggert, *Securing the Past*, 24.
37. *The Ecclesiologist*, 1: 65 (1842), 6. Quoted in James F. White, *The Cambridge Movement: The Ecclesiologists and the Gothic Revival* (Cambridge: Cambridge University Press, 1962), 159–60.
38. George Gilbert Scott practised restoration according to the function of the building. Churches, the function of which is to house God, had to be restored to the best way possible. Less restrictive restoration principles applied to other old structures that do not have their original function and refer to previous civilisations.
39. Jokilehto, *A History of Architectural Conservation*, 162.
40. John Ruskin, *The Seven Lamps of Architecture* (Elibron Classics, 2005 [1849]), 353.
41. Ruskin, *The Seven Lamps of Architecture*, 355.
42. Ruskin, *The Seven Lamps of Architecture*, 339.
43. Ruskin, *The Seven Lamps of Architecture*, 358.
44. Jokilehto, *A History of Architectural Conservation*, 175.
45. Jonathan Rée, 'Auto-Icons' in *Conservation: Principles, Dilemmas and Uncomfortable Truths*, ed. Alison Richmond and Alison Bracker (London: Elsevier/Victoria & Albert Museum, 2009), 1–2.
46. Jokilehto, *A History of Architectural Conservation*, 156.
47. Stubbs, *Time Honored*, 215.
48. Stubbs and Makas, *Architectural Conservation in Europe and the Americas*, 59.
49. Jokilehto, *A History of Architectural Conservation*, 174.
50. Salvador Muñoz Viñas, *Contemporary Theory of Conservation* (Oxford and Burlington, MA: Elsevier Butterworth-Heinemann, 2005), xii.
51. Nicholas Stanley-Price, Mansfield Kirby Talley and Alessandra Melucco Vaccaro, eds, *Historical and Philosophical Issues in the Conservation of Cultural Heritage* (Los Angeles: Getty Conservation Institute, 1996), 326–31.
52. Stanley-Price, Talley and Melucco Vaccaro, eds, *Historical and Philosophical Issues in the Conservation of Cultural Heritage*, xiii.
53. Manuel Martín Hernández, 'Architecture from Architecture: Encounters between Conservation and Restoration', *Future Anterior*, 4: 2 (Winter 2007), 62.

54. Alois Riegl, 'The Modern Cult of Monuments: Its Character and Its Origin' [1903], *Oppositions*, 25 (Fall 1982), 26.
55. Riegl, 'The Modern Cult of Monuments', 34.
56. Riegl, 'The Modern Cult of Monuments', 32–3.
57. Riegl, 'The Modern Cult of Monuments', 34–8.
58. Riegl, 'The Modern Cult of Monuments', 39.
59. The Charter of Krakow 2000, Principles for Conservation and Restoration of Built Heritage, http://smartheritage.com/wp-content/uploads/2015/03/KRAKOV-CHARTER-2000.pdf, accessed 21 August 2018.
60. The Charter of Krakow 2000.
61. Manuel Martín Hernández, 'Architecture from Architecture: Encounters Between Conservation and Restoration', *Future Anterior*, 4: 2 (Winter 2007), 66–7.
62. The Charter of Krakow 2000.
63. The Charter of Krakow 2000.
64. English Heritage, 'Conservation Principles, Policies and Guidance', 67, principles 160 and 164.
65. Historic England, 'Definition: Heritage Asset', https://historicengland.org.uk/advice/hpg/hpr-definitions/h/536274/, accessed 21 August 2018.
66. 'Identification and Designation of Heritage Assets', https://historicengland.org.uk/advice/hpg/has/, accessed 21 August 2018.
67. 'Listed Buildings', https://historicengland.org.uk/advice/hpg/has/listed-buildings/, accessed 21 August 2018.
68. Jorge Otero-Pialos, 'Editorial: Creative Agents', *Future Anterior*, 3: 1 (Summer 2006), ii–vii.
69. Gaston Bachelard, *The Dialectic of Duration*, translated by Mary McAllester (Manchester: Clinamen, 2000), 28.
70. Bachelard, *The Dialectic of Duration*, 30.
71. Bachelard, *The Dialectic of Duration*, 30.
72. Bachelard, *The Dialectic of Duration*, 29.
73. Conrad Russell, 'Fictive Time: Bachelard on Memory, Duration and Consciousness', *Kronoscope*, 5: 1 (2005), 7.
74. Bachelard, *The Dialectic of Duration*, 137–8.
75. Bachelard, *The Dialectic of Duration*, 138.
76. Russell, 'Fictive Time', 8.
77. Bachelard, *The Dialectic of Duration*, 64.
78. Gaston Bachelard, *The Poetics of Space*, translated by Maria Jolas, with foreword by John R. Stilgoe (Boston: Beacon Press, 1994), 9.
79. Bachelard, *The Dialectic of Duration*, 62.
80. Bachelard, *The Dialectic of Duration*, 61.
81. Conrad, 'Fictive Time', 11.
82. As opposed to *fictitious*, which implies something imaginary or fabricated.
83. Definition taken from the *Concise Oxford English Dictionary*.
84. Edward Kaplan, 'Gaston Bachelard's Philosophy of Imagination: An Introduction', *Philosophy and Phenomenological Research*, 33: 1 (September 1972), 2–3.
85. Caroline Joan S. Picart, 'Metaphysics in Gaston Bachelard's "Reverie"', *Human Studies*, 20: 1 (January 1997), 63.
86. Gaston Bachelard, *Water and Dreams: An Essay on the Imagination of Matter*, translated by Edith Farrell (Dallas: Pegasus Foundation, 1983 [1942]), viii.
87. Bachelard, *Water and Dreams*, viii.
88. Bachelard, *Water and Dreams*, 1.
89. Bachelard, *Water and Dreams*, 8.
90. Bachelard, *The Poetics of Space*, xxvii.
91. Bachelard, *The Poetics of Space*, 107.
92. Picart, 'Metaphysics in Gaston Bachelard's "Reverie"', 66.
93. Picart, 'Metaphysics in Gaston Bachelard's "Reverie"', 70.
94. Bachelard, *The Poetics of Space*, 9.
95. Frédéric Lefevre, 'Une heure avec Maurice Maeterlinck', *Les Nouvelles Littéraires* (7 April 1928), 8, quoted in R. C. Grogin, *The Bergsonian Controversy in France 1900–1914* (Calgary: University of Calgary Press, 1988), 61, quoted in *The New Bergson*, ed. John Mullarkey (Manchester and New York: Manchester University Press, 1999), xvii.
96. Henri Bergson, *Time and Free Will* (London: Elibron Classics, 2005 [1889]), 97.

97. Henri Bergson, *Duration and Simultaneity: Bergson and the Einsteinian Universe*, translated by Leon Jacobson (Manchester: Clinamen Press, 1999), 34–5.
98. Gilles Deleuze, *Bergsonism* (New York: Zone, 1988), 13.
99. Deleuze, *Bergsonism*, 22.
100. The influence of Bergson's method of intuition as a philosophical methodology that requires the split of a composite into its tendencies can be also traced in Walter Benjamin's 1917 essay 'Painting, or Signs and Marks'. In this essay, Benjamin explains that a mark differs from a sign in that the latter is printed on a surface whereas the former emerges from it. Also, a mark usually appears on living beings, yet it can manifest itself on lifeless objects too. For Benjamin, the extraordinary quality of a mark emerges when it comes to reveal a characteristic inherent in living beings, such as guilt or innocence. With blushing, for instance, which is a sign most commonly related to guilt, this reaction tends to uncover basic components of one's own personality. Therefore the line traced on a piece of paper in Bergson's example can be considered as a *mark,* which allows one to reveal the composite's (hidden) tendencies.
101. For Bergson, a person's physical state reflects his entire personality. In *An Introduction to Metaphysics*, he claims that the philosophers' attempt to reconstruct somebody's personality from his physical states leads the discipline to an unavoidable confusion. One's feeling contains virtually the entirety of his past and present. If a psychologist for instance wants to detach a mental state in order to examine it as a separate entity in itself, he must use a method of abstraction or of analysis, thus neglecting the full range of shades of the examined personality. This way '[they] mistake partial notations for real parts, thus confusing the point of view of analysis and of intuition, of science and metaphysics'. Henri Bergson, *An Introduction to Metaphysics*, translated by T. E. Hulme (Indianapolis: Hackett, 1999), 34. For this reason every composite, like memory and experience, should be divided into its initial articulations through the method of intuition.
102. Deleuze, *Bergsonism*, 33.
103. Henri Bergson, *Creative Evolution*, translated by A. Mitchell (New York: Henry Holt, 1911), 94.
104. Bergson, *Creative Evolution*, 89.
105. Bergson, *Creative Evolution*, 89.
106. James Burton, 'Bergson's Non-Archival Theory of Memory', *Memory Studies*, 1: 3 (September 2008), 323.
107. Henri Bergson, *Matter and Memory*, translated by M. N. Paul and S. Palmer (New York: Dover, 2004 [1896]), 72–3.
108. Bergson, *Matter and Memory*, 72.
109. Bergson, *Matter and Memory*, 5.
110. Bergson, *Matter and Memory*, 8.
111. Bergson, *Matter and Memory*, 34.
112. Bergson, *Matter and Memory*, 89–90.
113. Bergson, *Matter and Memory*, 92.
114. Deleuze, *Bergsonism*, 59–61.
115. Burton, 'Bergson's Non-Archival Theory of Memory', 327.
116. Burton, 'Bergson's Non-Archival Theory of Memory', 326.
117. John Mullarkey, *Bergson and Philosophy* (Edinburgh: Edinburgh University Press, 1999), 51.
118. Burton, 'Bergson's Non-Archival Theory of Memory', 330.

5
Architecture and Fire

> The beginning of association among human beings, their meeting and living together, thus came into being because of the discovery of fire.
>
> Vitruvius, *De Architectura*, Book II (*c*. 30–20 BC)

There is a shared assumption among architectural historians and theorists that the primitive hut[1] was erected around fire. This suggests that fire was once central to architectural space, which agrees with the general notions of fire's centrality in early human life and thought. It also reveals that architectural evolution has a starting point, the fire of the primitive hut. Before fire there is a dark empty space associated with oblivion; with fire comes architecture, and thus architecture's first memory.

The Roman architect and writer Vitruvius in *De Architectura* (*c*. 30–20 BC) describes how human life changed as soon as fire was invented. Observing how fires started accidentally through the rubbing of tree branches, and appreciating the thermal comfort produced by the flames, primitive man gradually learnt how to ignite fire and prolong the burning process. This accidental discovery of fire, Vitruvius suggests, is what 'originally gave rise to the coming together of men, to the deliberative assembly, and to social intercourse'.[2]

Tracing the origins of architecture to the primitive hut, the German architect and writer Gottfried Semper (1803–1879) considers that the fire of the hearth is the 'first and most important, the moral element of architecture'.[3] The hearth is in the centre of the primitive hut and around it three distinct elements emerge: the roof, the enclosure and the mound or foundation. Semper argues that 'throughout all phases of society the hearth formed the sacred focus around which the whole took order and shape'.[4] The first signs of human settlements coincide with the lighting of fire for survival, warmth and the preparation of food. Following Vitruvius, the fire of the hearth acts not only as an assembly point but also as the point where alliances were formed for the first time and where religious rituals were practised.

The element of fire also holds a unique position in myths of origin, and here it is similarly associated with architecture and construction. In ancient Greece, there was a goddess dedicated to the hearth and the home called Hestia. Her flames burnt in the *prytaneion* (town hall), which symbolised the centre of city life. In ancient Rome, the goddess of the hearth was called Vesta, and she ruled both the fire of the household and the civic hearth of the city. According to the architectural historian and critic Joseph Rykwert, ancient Rome developed around the *mundus*, a hole in the ground that symbolised the mouth of the Underworld. The mundus indicated the symbolic centre of the city[5] and its opening was sealed with a stone altar, which burnt a constant fire. Every subsequent Roman city featured its own mundus and burning fire taken from the initial altar in Rome.[6]

Fire and thermal comfort

Historical enquiry and the analysis of relevant myths reveals that fire occupies a distinct position in architecture. It is found at the beginning of architectural evolution, occupying a central position in space. The significance attached to the element since its invention suggests further that primitive man immediately assessed the advantages of fire and appreciated them immensely, largely because of the thermal comfort that the flames provide. Thermal comfort relates to the satisfaction that the body receives in an environment and is perceived through a bodily function called thermal sense, which is the ability to distinguish differences of temperature. While people have a sense for warmth and coolness, thermal sense is not included in the five traditionally recognised senses. In most cases it is considered as part of the sense of touch, as through touching one can identify the temperature of an object. The philosopher F. G. Asenjo argues that the 'thermal sense is most anthropocentric of all, for its constant reference to skin frames each perception more definitely around the self than do pressure or vibration. The thermal sense certainly places our body at the center of the universe.'[7]

The ability of our body to sense if it is getting cooler or warmer is attributed to a function of the nervous system. The nervous system is programmed to react to changes in the environment rather than to steady states. When we enter a warm room directly from a cold environment, we first feel heat through our thermal sensors until the body adapts to the new conditions. Therefore the perception of differences in temperature occurs through a comparative process that unfolds in time.[8] Yet the thermal sensors are not distance receptors. We cannot perceive the heat of a room until we find ourselves inside it. For this reason we employ other senses to gives us clues in advance. Effectively, the more senses we employ to experience our environment, the more complete our perception of it becomes. Referring to the impact of fire on perception, the architect and researcher Lisa Heschong explains that fire offers a totality of sensory stimulation.

> The fire gives a flickering and glowing light, ever moving, ever changing. It crackles and hisses and fills the room with the smells of smoke and wood and perhaps even food. It penetrates us with its warmth. Every sense is stimulated and all of their associated modes of perception, such as memory and an awareness of time, are also brought into play, focused on the one experience of the fire. Together they create such an intense feeling of reality, of the 'here and nowness' of the moment, that the fire becomes completely captivating. We are likely to feel that we could spend hours mesmerised by it.[9]

Thermal sense is absorbed and integrated into our culture in the form of an unconscious everyday practice, which is the result of thermal associations that we develop in the environments we inhabit. On hot summer days, for instance, we tend to retreat to cooler parts of a building that are not exposed to the hot rays of the sun. The fondness we develop towards certain rooms of the house with regard to the thermal comfort they provide is often evident in the attention to their decoration. For example, during the Middle Ages the hand-woven carpets and tapestries hung on walls superseded their basic function of keeping the cold out of the house. They became objects of affection and were regarded one of the highest forms of art.[10]

Philippe Rahm's *Digestible Gulf Stream*

Taking into consideration the necessity and fondness for thermal comfort and turning it into a design objective, the contemporary Swiss architect Philippe Rahm developed an experimental architectural installation called *Digestible Gulf Stream*, which was showcased at the Venice Architecture Biennale in 2008 (fig. 5.1). For Rahm,

Figure 5.1 Site-specific installation by Philippe Rahm Architectes in the Corderie at the Arsenale, Venice Architectural Biennale, 2008. The inhabitants experience a small-scale gulf stream effect, which aims at increasing their thermal comfort. © Philippe Rahm Architectes.

architecture 'should no longer build spaces, but rather create temperatures and atmospheres'.[11] His installation consisted of two horizontal metal planes placed on different heights. The temperature of the lower one was 28°C whereas the temperature of the upper one was 12°C. The difference of temperature between the two planes creates a small-scale gulf stream phenomenon, as there is a constant flow of warm air through convection. This way Rahm exposes an invisible meteorological parameter, namely temperature, and manipulates it to define a 'fluid, airy, atmospheric space'.[12]

The hearth of the early house

Between the twelfth and seventeenth centuries, the majority of houses in Europe, North America and the Middle East were heated by an open fire. This indoor hearth burnt in the middle of rooms and consisted of a shallow pit lined with stones. A higher stone (reredos) with a hollow surface was placed behind it acting as a fireback. The heat and smoke escaped from a hole on the gable roof that was located directly above the fire. The thermal radiation produced by the flames was beneficial only when people sat directly in front of the fire, whereas the building itself was only superficially heated, as hot air escaped through the hole. Examples of this primitive form of fireplace, which are possibly the earliest to survive, can be found on the Orkney Islands and the Shetlands in Scotland.[13] The rooms without a hearth were often heated using transportable charcoal braziers. For safety reasons, these braziers were commonly started up outdoors. As soon as the charcoal was glowing, the braziers were either wheeled in or carried to the rooms that needed warmth.[14] A typical example of a central hearth in the United Kingdom can be found in the Baron's Hall at Penshurst Place, a mansion built in 1341 in Penshurst, Kent. In the middle of the Hall there is an octagonal hearth. There is no hole in the roof for the smoke to escape but the gable windows have unglazed round openings.[15]

The first step towards the invention of chimneys was made in the late thirteenth century when the hole of the gable roof was covered with louvres to keep the rainwater out.[16] The chimney was introduced relatively late and even then it was only used as a means to direct smoke out of the house. Its function to create draughts had yet to be invented.[17] The earliest written reference to chimneys stretches back to 1347, when an earthquake in Venice destroyed a big number of them (*molti camini*).[18] The construction of the first chimneys led to many disastrous fires, as they were built out of timber. Gradually the knowledge acquired from previous accidents pointed towards the use of non-combustible materials such as stone and brick. These materials were not only safer but also more energy efficient, as the thermal radiation from the fire was absorbed by the material of the chimney and was then reradiated to heat up the room. Based on this principle, medieval castles featured thicker walls opposite fireplaces, so that they could absorb the heat and return it back into the room as soon as the fire was out.[19] The Norman castle (*c*. 1100) at Hedingham (fig. 5.2) in Essex, England for instance features fireplaces on three levels. The walls of the castle are thick on purpose so that they can store heat and radiate it back to the building. As the technology of chimneys had not developed by then, the flues of the castle's fireplaces were only used to direct smoke out of the building.[20]

By the sixteenth century, people's demands for thermal comfort had increased and the great house no longer needed any fortification. Therefore the hearth moved from the centre of the house to a recess in the external wall to create what we know today as a fireplace.[21]

Figure 5.2 Cross-section of Norman castle at Hedingham in Essex (*c.* 1100), revealing the thickness of walls. From John Alfred Gotch, *The Growth of the English House: A Short History of its Architectural Development from 1100 to 1800* (London: Batsford, 1909) © The British Library Board.

History of matches

Fire proliferated in architectural space as soon as it became readily available in every house. A revolutionary discovery that occurred accidentally in Hamburg, Germany, in 1669 was to change the course of fire's proliferation within the urban landscape.[22] An experiment on urine conducted by an alchemist called Henning Brandt in his attempt to find the philosopher's stone resulted instead in the discovery of phosphorus,[23] a substance described by the science writer John Emsley as 'the devil's element'.[24] The use of phosphorus in matches gave freedom to the individual to start a fire at home.

Before the invention of phosphorus matches, there were many attempts to create self-igniting ones. The most common practice involved wooden sticks with one end dipped in molten sulphur. The contact of sulphur with a hot surface or a spark created by the friction of flint on metal ignited the match. Sulphur matches were present in most eighteenth-century kitchens. Chlorate matches were another invention of the same period; they contained potassium chlorate, sugar and gum, and ignited as a result of the chemical reaction with sulphuric acid. They were produced for the first time in France in 1805. In 1810, Henry Berry invented a similar product in London.[25] This was a box that was half filled with chlorate matches, while the other half contained a bottle with a piece of asbestos wool soaked in sulphuric acid. The danger of these matches lay in the necessity of transporting sulphuric acid for ignition. In 1828, Samuel Jones provided a solution with the Promethean match.[26] This match consisted of a glass tube that contained sulphuric acid and was wrapped with paper. One end of the paper was dipped in potassium chlorate, sulphur and sugar. The paper ignited as soon as the tube was cut and the acid was released from inside. Attempts to use phosphorus on matches were also made, but the cost of producing it as well as its uncontrolled flammability made it rather an undesirable alternative.

The turning point in the history of matches occurred once again accidentally, in 1825. John Walker was a pharmacist who ran a shop in Stockton-on-Tees. He was appointed by one of his clients to create a mix of potassium chlorate, antimony and gum when a sample of it fell on the hearth of his shop. Later, when the mix was dry and Walker trod on it, he observed that it ignited. This accidental discovery led to the invention of friction matches, which were sold for the first time by Walker in 1827.[27] Samuel Jones, inspired by a lecture on this invention given by Michael Faraday at the Royal Institution in London in 1828, produced his own version of self-igniting matches called Lucifers, a name that was used for all self-igniting matches that followed.[28] A discovery not long after, by a Swedish mineralogist who sourced calcium phosphate extracted from bones, allowed the cheap production of phosphorus. As a result, a small quantity of this element was introduced to Lucifers, which led to the sudden advancement of the match industry. Nonetheless, the new phosphorus Lucifers entailed further risks. The flammability of the wood, or the 'afterglow' of the match after the fire was out, led to many unpredictable domestic fires. Moreover, the element was highly toxic and was easily ignitable. Anton Schrötter von Kristelli in Austria provided a solution to these dangers with the invention of red phosphorus,[29] after which the match industry developed dramatically on an international level.[30]

The invention of matches marks the moment in history when fire became easily available in every house. At the turn of the twentieth century, following the introduction of gas and later of electricity, fire was omnipresent in both domestic and industrial buildings, causing a sudden and radical reconfiguration of architecture and urbanism.

The emergence of the central heating system

The escalating demand for thermal comfort led progressively to the invention of sophisticated heating systems and equipment. The most popular heating method today is the central heating system, an invention commonly attributed to the ancient Roman hypocaust, a heating system that was used in Roman baths and the mansions of wealthy people. More recent studies in archaeology, anthropology and architecture indicate that heating technology in certain areas of Asia preceded the ancient Roman system by thousands of years.[31] Excavations in northern China and northern Korea brought to the surface three early radiant heating systems called *kang, dikang* and *ondol*, some of which have been dated between 5300 and 4800 BC. The *kang* was a heating system that involved a raised heated surface destined for sitting or sleeping. The *dikang* was a similar system that also included a fireplace. Hot gases emitted through combustion in the fireplace travelled through flues within floors or walls, which radiated heat to the room.[32] The *ondol* refers to a heated floor system used in Korea for both sleeping and sitting.[33]

Figure 5.3 Interior of a greenhouse heated by Musgrave's iron slow combustion stove. From Shirley Hibberd, *The Amateur's Greenhouse and Conservatory: A Handy Guide to the Construction and Management of Planthouses* (London, 1873) © The British Library Board.

Only much later was the Roman hypocaust invented in Europe. In the hypocaust, water was heated by a central fire and circulated in flues under the floor or through the walls of a room. The hot water was first piped to the hottest bath chamber, the *caldarium*, and then to the *tepidarium*, a warm pool. Then, as the water was cold again, it was channelled to the *frigidarium*, which was a cold pool. The smoke and gases released by the combustion travelled under the *caldarium* in the hypocaust and up through flues concealed in walls that vented to the open air.[34] The hypocaust is often believed to have inspired the heating of greenhouses in Europe from the eighteenth century onwards.[35] Solar power was often insufficient for warming greenhouses and therefore a burning stove was placed centrally inside them (fig. 5.3).

The intention of the stove was to heat the greenhouse further, but this solution did not achieve uniform distribution of heat. Hence the relocation of the stove in a space adjacent to the greenhouse or in the basement, where it boiled water that was carried along the perimeter through a hidden system of pipes (figs. 5.4 and 5.5). This solution was more successful as heat was now distributed uniformly throughout the entirety of the greenhouse. This technology of heating greenhouses inspired in turn the emergence of the central heating system.[36]

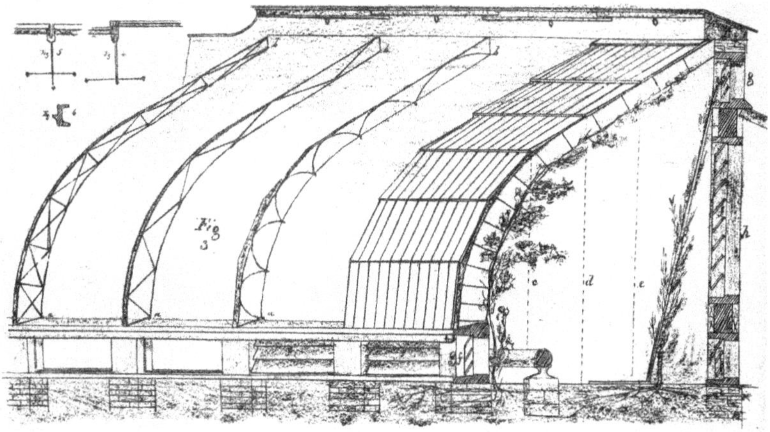

Figure 5.4 Section of a greenhouse that is destined for the cultivation of vines. The greenhouse is heated by steam with smoke flues installed in the back wall. From J. C. Loudon, *Remarks on the Construction of Hothouses* (London, 1817) © The British Library Board.

The architectural historian Robert Bruegmann attributes the emergence of the three main central heating systems – hot air, hot water and steam – to technological advancements that took place mainly in Great Britain between the late eighteenth and early nineteenth centuries. The hot air heating system involves the distribution of hot air produced by a closed stove and its development is primarily attributed to William Strutt. Strutt applied this technology widely to the heating of mills, as in the case of a fireproof mill in Derby (1792–3) that was heated through a single hot air cockle stove (furnace).[37] A cockle stove was also used for the heating of the Derbyshire Infirmary that was designed and installed between 1806 and 1810.

Hot air central heating was usually the preferred option, as it was a financially viable system and its installation did not demand high engineering skills. This system was particularly popular during the first half of the nineteenth century and it was used both in large public buildings, like the heating system in the House of Commons designed by the Marquis J. B. M. F. de Chabannes, and in smaller private ones.[38]

Steam central heating developed in parallel with hot air heating. The earliest system was designed by Matthew Boulton and James Watt in the end of the eighteenth century and was installed on the Twist Mill at Salford. The earliest multi-storey steam heating system was installed in a mill at Dornoch, Scotland, in 1799 and was designed by Neil Snodgrass.[39]

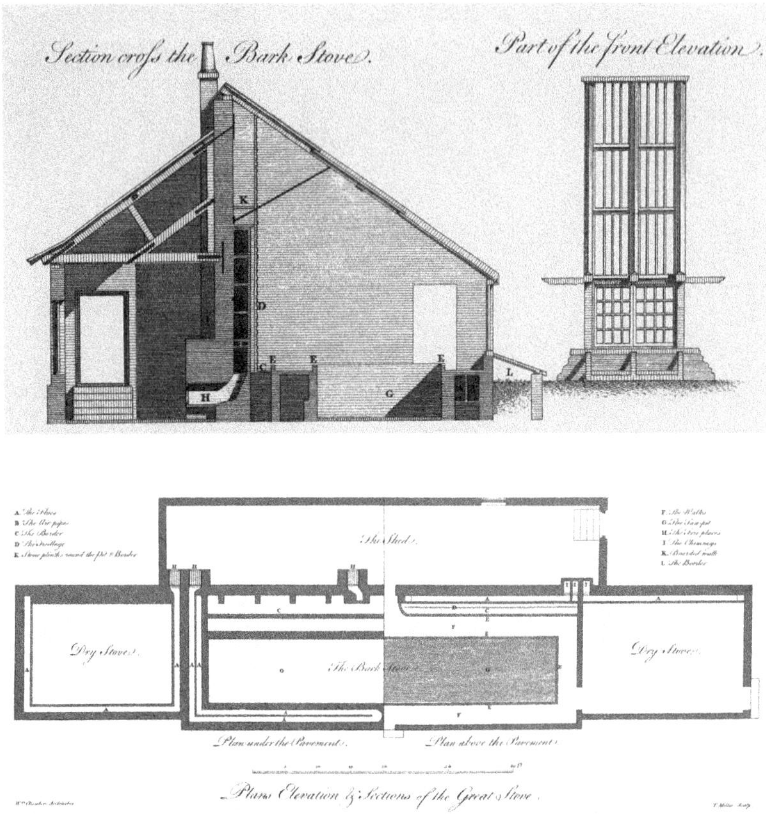

Figure 5.5a, b Plan, elevation and section of the Great Stove in Kew Gardens designed by Sir William Chambers. The intention of this stove was to house 'the amplest and best collection of curious Plants, in Europe'. In the middle of the Great Stove there is a large bark-stove and on either side there is one smaller dry-stove. The bark-stove is heated by four furnaces that warm the air carried in the flues under the pavement and behind the back wall. From Sir William Chambers, *Plans, Elevations, Sections, and Perspective Views of the Gardens and Buildings at Kew in Surrey* (London, 1763; repr. Farnborough: Gregg Press, 1966). © The British Library Board.

According to a reference that was made to Snodgrass's installation in 1806 in *Transactions of the Society, Instituted at London, for the Encouragement of Arts, Manufactures, and Commerce* (see also fig. 5.7), Snodgrass

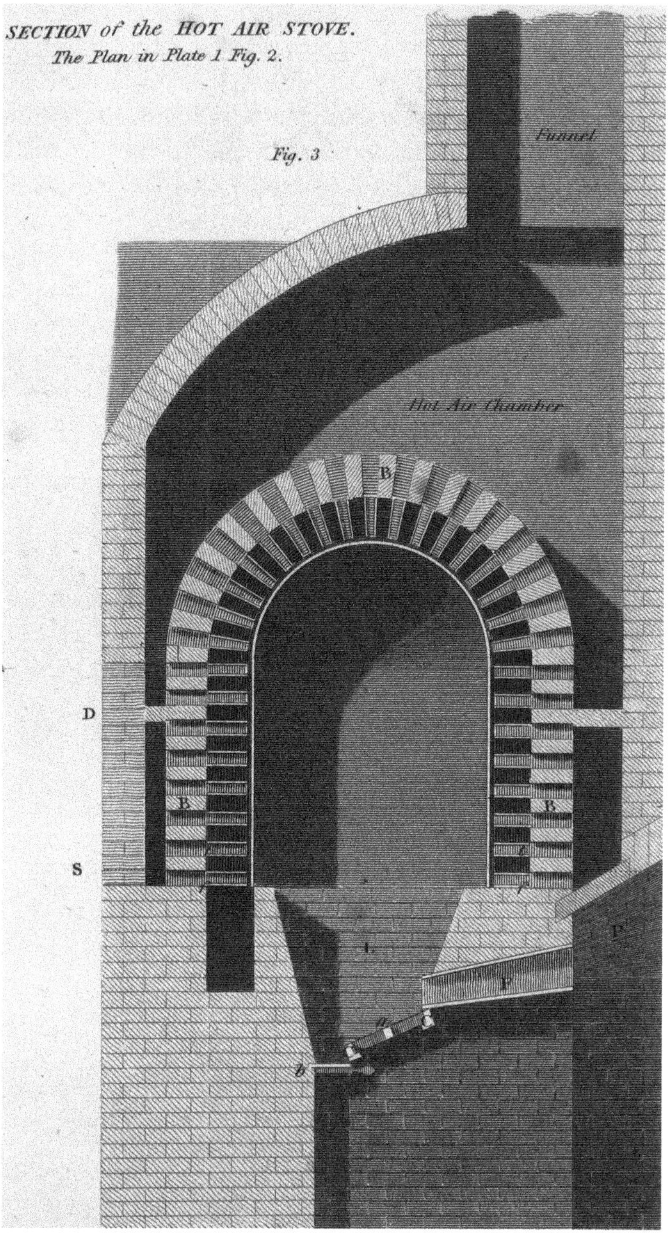

Figure 5.6 Strutt's 'cockle' stove at Derbyshire Infirmary, 1806–10. From Charles Sylvester, *The Philosophy of Domestic Economy, As Exemplified in the Mode of Warming, Ventilating, Washing, Drying, and Cooking … Adopted in the Derbyshire General Infirmary, etc.* (Nottingham, 1819). © The British Library Board.

was led to consider of a cheap method of heating the mill, as he had learnt that fuel was scarce and dear in the country in which the mill was situated. It was evident that none of the methods which he had seen practised could be applied, but at an enormous expense; and his experience had pointed out to him important defects and inconveniences in them all. Having observed a mode of drying muslins by wrapping them round hollow metal cylinders, filled with steam, practised at the bleach-fields near Glasgow, it occurred to him, that by means of a proper apparatus, steam might be applied to heat a cotton mill, or any other large manufactory. It was evident that this not only would be [an] economical mode of producing heat in large works, so far as fuel was concerned, but that it would prevent the danger of fire, to which such works, when heated in the usual manner, are much exposed.[40]

The proposed scheme included a boiler located outside the mill that produced steam to be circulated throughout the building in steam pipes. These pipes ran close to the walls and discharged at the top of the building. The steam heating system experienced great success during the second half of the nineteenth century and was mainly applied to large-scale buildings.[41]

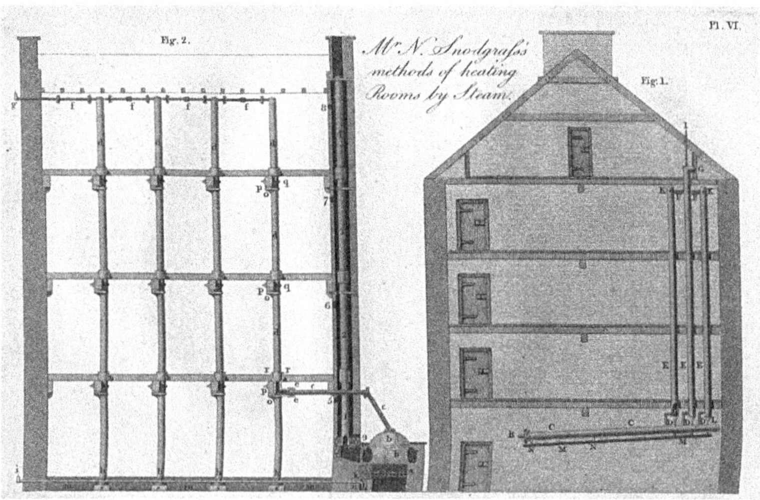

Figure 5.7 Snodgrass's proposal for using columns as steam heating pipes, 1806. From *Transactions of the Society, Instituted at London, for the Encouragement of Arts, Manufactures, and Commerce*, vol. 24. © The British Library Board.

Hot water central heating was developed in the 1840s and its emergence is often attributed to Jean Simon Bonnemain, a French mechanical engineer who invented a hot water system for chicken incubators between 1778 and 1793.[42] The initial versions of these heating systems had low pressure so the water was circulated in pipes with a large diameter. Following the invention of high-pressure hot water systems by Jacob Perkins and Angier March, the large pipes were replaced by smaller ones that could be hidden away within walls, floors and ceilings. A high-pressure hot water system was installed in Sir John Soane's Museum and House in 1891, as revealed by Richardson, one of Soane's pupils. This system proved particularly effective, as the pipes were concealed from visitors and the multiple rooms of the building were heated efficiently.[43]

Nineteenth-century architects compromised their designs following requirements imposed by the advancements of central heating systems. For instance, the Wakefield Asylum (built 1816–18) designed by Watson and Pritchett featured circular stairwells that were located at the intersections of four wings (see fig. 5.8). A cockle stove burnt at the bottom of each stairwell, which heated up air that rose up into the well

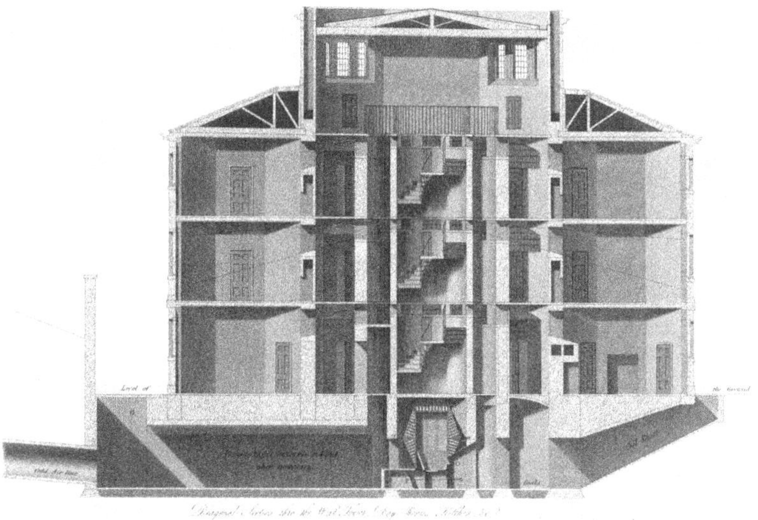

Figure 5.8 Cross-section of the Wakefield Asylum designed by Watson and Pritchett, built 1816–18. From Watson and Pritchett Architects, *Plans, Elevations, Sections, and Description of the Pauper Lunatic Asylum … at Wakefield* (York, 1819). © The British Library Board.

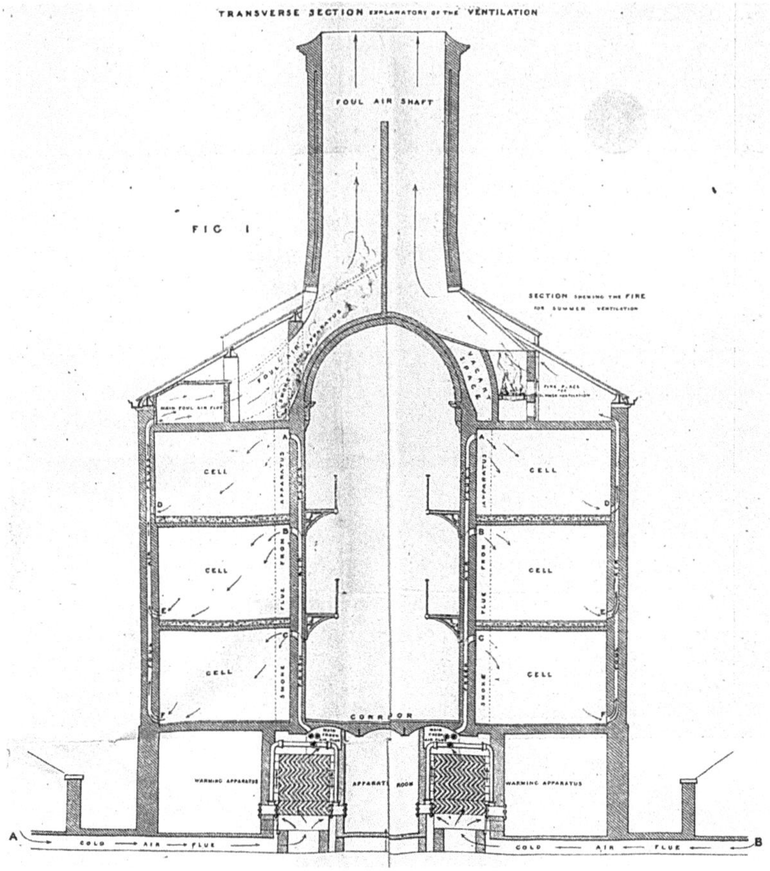

Figure 5.9 Cross-section of Pentonville Prison in London, built 1841–2. From Surveyor-General of Prisons, 'Report of the Surveyor-General of Prisons on the Construction, Ventilation, and Details of Pentonville Prison' (London, 1844). © The British Library Board.

and spread through transom windows into the wards. The same windows were also used to provide uninterrupted views from the stairwells into the wards for surveillance reasons.[44]

The heating of Pentonville Prison in London (built 1841–2) was achieved through a carefully devised central heating system (fig. 5.9). Basement boilers heated air that travelled through ducts concealed within the thick corridor walls. Each cell had a warm air inlet at the top and an air outlet at the bottom that discharged through the prison's external wall. This design allowed each cell to be independently heated and ventilated.[45]

Reflecting on architecture's adaptability with regard to the advancements of central heating systems, the architect Luis Fernández-Galiano traces similarities with Jeremy Bentham's Panopticon as an example of emerging modernity. Whereas the design of the Panopticon was conceived as a way to allow the warden, who was located centrally, to observe the prisoners without obstructions, central heating systems 'guaranteed homogenous "artificial warmth" that would safeguard the health of the bodies in the same way that the vigilant eye tried to reform the ways and safeguard the health of the souls. The Panopticon was therefore a panthermicon.'[46]

The evolution of heating systems demonstrates that fire starts to gradually distance itself visually from architecture. Processes of combustion still take place within the building but are hidden away from the inhabitants. Basement boilers occupy their own physical space and distribute heat silently through an invisible system of valves, drains and pipes.

Fire and modernism

With the advent of the modern movement, fire and architecture are irreversibly divorced. This moment marks a sudden break with history, and the search for an *architecture of purity* expressed through amnesia and thermal silence. In *Walls Have Feelings*, Katherine Shonfield suggests that this architecture of purity reflects 'the idea of *the defence of the border*',[47] according to which borders define and contain only whatever is pure. Modernist architecture is thus pure, for its borders are clearly distinguished and defined. The heating systems of modernist buildings are equally pure, as the spaces they occupy are clearly defined. The boiler room serves the building through a highly sophisticated network of pipes concealed within walls and ceilings. Modern heating systems occupy their own space and are visually disconnected from the inhabitant: a space for pipes and a space for people.

Commenting on the modernisation of domestic space, the curator and art historian Helen Molesworth refers to the architectural historian Sigfried Giedion, who in his 1948 book *Mechanization Takes Command* describes how the twentieth-century home 'developed a "mechanical core" around the kitchen and the bathroom … the home was [also] re-structured around water availability in order to incorporate standardised plumbing, was wired for electricity, saw the rise of centralised heating and experienced the introduction of the phone'.[48] Molesworth adds to Giedion's description that the overall experience of the mechanisation of domestic space was about *absorption*. The home absorbed and reflected aspects of industry; therefore the 'connected counter-top that ran around the perimeter of the kitchen, allowed for an easy incorporation of new tools and technologies to aid in the efficiency of household labor'.[49] Reflecting on Molesworth's statement, I argue that a similar process of absorption took place with heating systems, which has in the long run altered our perception of thermal comfort. The early heating systems operated with a stove external to the building that produced hot air, steam or water and distributed heat through a visible network of pipes or ducts. Later, the boiler-radiator heating system was *absorbed* into the structure of the building where concealed pipes ran between floor slabs and floor finishes. We can still sense the heat radiated through them, but the lack of direct visual contact leads to an abstract, decentralised appreciation.

Fire and sustainable architecture

In a period of global environmental awareness, the relationship between architecture and fire is subject to further alterations. Amongst various demands, sustainable agendas aim to reduce drastically, if not to end, combustion in buildings. Advancement in science and technology has made possible the creation of habitable environments that can overcome the laws of nature with the simple flick of a switch. This is achievable by consuming high levels of energy, which are typically sourced from fossil fuels such as oil, natural gas and coal. Fossil fuels are a natural source and their supply is finite. Their consumption entails economic, social and environmental implications. This consumption is responsible for global warming, air and water pollution, and land damage from mining and transportation, and means nations are often dependent on foreign energy sources. The catastrophic impact on the environment begins when fuels are removed from their source and lasts until the point when they are consumed for energy supply.[50]

Since the International Conference on Climate Change held in Tokyo in 1997, governments worldwide have agreed on reducing their environmental impact. Architecture's response has been the implementation of a sustainable design approach, which is a 'revised conceptualisation of architecture in response to a myriad of contemporary concerns about the effects of human activity'.[51] Compared to traditional design approaches, sustainable design considers buildings as systems within specific environments (man-made or natural ones) that can be affected by numerous variables. One of the difficulties with this is that there is no limit to the number of variables; therefore sustainable design is to a certain degree 'prognostic and anticipatory'.[52] The variables that are always taken into consideration are the site location, the existing vegetation and landscaping, the building orientation, the building envelope, the external glazed surfaces, the construction of walls and roofs, the insulation, the natural ventilation and lighting possibilities, and the construction materials. These variables are crucial to sustainable design as they determine the levels of thermal comfort in a building and thus also the levels of energy consumption.[53]

Sustainable design targets the reduction or, if possible, even the abolishment of combustion. Green buildings employ alternative energy sources, namely solar, wind and hydro energy as well as biomass, which are called Sustainable Energy Sources (SES) or Renewable Energy Sources (RES). Energy is captured from existing natural flows (the sun, wind, flowing water), from biological processes and geothermal heat.

The design of buildings in relation to the sun has been discussed since ancient times. For instance, in *Memorabilia* (Book III, Chapter VIII) the ancient Greek historian Xenophon (430–354 BC) cites Socrates, who questions the orientation of buildings according to sunlight and puts forward basic sustainable design principles:

> In houses looking toward the south, does the sun shine into the inner rooms in winter, while in the summer when it travels over us and the roofs, it provides shade? If then it is noble that these things come to be thus, one should build a house higher on the southern side, so that the winter sun won't be shut out, and closer to the ground on the northern side, so that the cold winds won't burst in. In sum, the most pleasant and most noble dwelling would plausibly be the one where one would have in every season the most pleasant refuge and where one would make one's belongings most secure.[54]

The French architect Le Corbusier took similar consideration of the sun path in the early stages of concept design, and this is also evident in his sketches.[55] Le Corbusier incorporated the *brise-soleil* system in his designs, which until then was a largely overlooked passive sun protection system traditionally used in North Africa. A typical *brise-soleil* system is composed of a horizontal projection sometimes incorporating a set of movable louvers, and is installed on the facades of buildings exposed to the sun. The *brise-soleil* system at Le Corbusier's Unité d'Habitation in Marseille (fig. 5.10) is designed in such a way that it does not allow the sunrays to touch the glass during the warmest hours of the day between the spring and autumn equinox. In the rest of the year, when the solar energy

Figure 5.10 Le Corbusier's *brise-soleil* system at the Unité d'Habitation in Marseille, built 1947–52. © Marie Dumas.

is beneficial and can warm up the interior of the building, the *brise-soleil* allows sunrays through.

Passive solar design deals with compositional arrangements, such as the admittance of energy through correctly oriented windows or other glazed openings, and the storage of energy in the thermal mass (i.e. floor slabs or walls), which is then distributed to internal spaces through natural convection and radiation. Passive solar design techniques can also be applied to existing buildings and can reduce energy consumption by up to 50 per cent. Passive design approaches do not demand the mediation of consuming devices such as pumps and fans to distribute energy throughout the building.[56]

In an active solar design, the energy of the sun is captured through the mediation of photovoltaic cell systems, which is then converted into electricity that can be used for any function in the building, such as heating and lighting. Active solar systems allow more flexibility in terms of design and installation, but as their functionality incorporates many moving parts, they have higher demand for maintenance. A common active solar system is the thermosyphon. This system includes a set of solar panels, which is used to heat a liquid that is present in a storage tank. The stored heat is then transmitted to a central heating device, which distributes the heat into radiators or underfloor heating systems.[57]

In areas with limited sunlight, sustainable buildings use other energy sources such as geothermal energy. This type of energy derives from radioactive decay found in the core of the planet. Exactly like solar energy, it can be used for the generation of electric current and the production of heat.[58] A building that makes use of geothermal energy is the Norddeutsche Landesbank in Hanover, Germany, designed by Behnisch and Partner in 2002. In this building, geothermal energy is used according to the season. During the hot months of the year, the heat absorbed by the ground is stored for later use during colder periods. The stored energy heats water that circulates in pipes integrated into the structure of the building. During winter, this function is reversed. The heat captured by the building is directed into the ground through the same pipework. The water that runs in the pipes is cooled with the use of the heat exchangers, and then circulates back into the building. The only non-sustainable energy that is needed for this heating/cooling system is electricity, which is consumed to run the circulation pump. In effect, the annual energy balance for the heat introduced and extracted from the ground is even.[59]

In sustainable buildings the combustion of materials for energy production ends. The spaces that fire once occupied, such as the boiler room, and associated building elements, such as fireplaces and chimneys, all become redundant. The memory of fire is projected outwards and is replaced by external processes. The heat produced by the sun and the earth replaces combustion, and architectural form is defined by an absence of the latter.

Architecture and thermodynamics

An energetic reading of architecture perceives a building as a composite of different energies that give life and order to inanimate matter. The building is no longer a mute autonomous object but rather a material organisation that emerges in time. Architecture's frozen essence defrosts, becomes warm, and the building itself can justifiably be perceived as an animate, habitable space. Considering architecture and the city as open thermodynamic systems means their material organisation can be linked with their information or memory. This association is confirmed by the ecologist Ramon Margalef, who states that nature has the capacity to store information. This is not restricted to the organic world. He illustrates this statement by referring to the gradual and random formation of a river's path, and the complexity of the earth's crust as it has developed throughout the years, which are both 'information-storing devices in the same manner that the genetic systems are'.[60] His main argument is that the information concealed in nature 'allows for a partial reconstruction of the past'.[61] Translating this into architecture, a building can be perceived as a store of useful memory. This memory is stored within materials, their properties and spatial arrangement, in technologies, in architectural and urban typologies, and stretches back to the construction of the primitive hut.

According to the second law of thermodynamics, as a system's entropy always increases and its ability to produce work decreases, energy from external sources must be introduced that can maintain the system's organisation. This suggests that with the passing of time, buildings can never sustain their initial form. They will always carry the impact of time and manifest their decay unless an external source of energy is introduced. This energy can reverse decay and preserve architecture's morphological organisation. This energy refers to the energy needed for the restoration of architecture.[62]

Architecture and excess of energy

Fire is an energy source that gives life to a building. There are occasions, however, when the energy consumed to create a building is the same energy that causes its destruction. Isolated fires burnt in incinerators create materials for the construction of big cities. Other isolated fires break out, causing catastrophic damage to large districts and whole cities. The French writer Georges Bataille, who is renowned for his views 'against architecture'[63] and against continuity of evolution,[64] comments on the energy of a living organism, explaining that

> in a situation determined by the play of energy on the surface of the globe, [the organism] ordinarily receives more energy than is necessary for maintaining life; the excess energy (wealth) can be used for the growth of a system (e.g., an organism); if the system can no longer grow, or if the excess cannot be completely absorbed in its growth, it must necessarily be lost without profit; it must be spent, willingly or not, gloriously or catastrophically.[65]

Excess of energy can thus be a reason for irretrievable damage. Based on this statement, Bataille draws parallels between the failure of absorbing the excess energy created by the growth of industrial activity and the outbreak of world wars. He explains that the excess of production that occurred due to the technical advancement of the early industrial period was difficult to sustain. As a result, this satiety brought about World War II, and the intensity of the war itself was a direct result of the size of this excess. Yet, Bataille adds, excess is not always necessarily destructive. Life manifested as growth and reproduction would not exist if an organism did not consume the surplus energy available in a productive way. A plant absorbs the energy offered by the sun and uses it for its own growth. It will occupy all the available space in any direction until growth reaches its limit. From this point onwards, if there is no more room for expansion, any further sum of energy absorbed becomes excessive.

> In a sense, life suffocates within limits that are too close; it aspires in manifold ways to an impossible growth; it releases a steady flow of excess resources, possibly involving large squanderings of energy. The limits of growth being reached, life, without being in a closed container, at least enters into ebullition: without exploding, its extreme exuberance pours out in a movement always bordering an explosion.[66]

Examples taken from nature that accurately illustrate Bataille's argument are the lodgepole pine in North America and the chamise or greasewood, a southern Californian shrub, which 'have evolved traits to stimulate fire'.[67]

Fire in architecture, like the energy of a living organism, can equally become excessive. Cities are created and are sometimes later destroyed by fire. The delicate equilibrium of the technologically sophisticated network that fire inhabits – visually disconnected from the dweller between walls and floors – can reach a momentum where energy's growth has no more room to expand. On these occasions it can take over the next adjoining space, which is space destined for habitation. Yet everyone is aware of its presence, and particularly of its vicious and catastrophic power. Our experience of it is not direct unless fire reveals itself accidentally. Nevertheless in most cases the experience we have is not of the actual event but rather of what fire leaves behind. This is the material end product of combustion and the memory of the accident.

Memory, materials and the evolution of technology

In *Theatres of Memory*, the historian Raphael Samuel claims that 'the philosophy of a building … is the spirit of the age incarnate, bearing the impress of the dreams and the dilemmas of its time'.[68] He discusses how brick as a building material carries with it notions of *pastness*, which give clues about previous political, economic and cultural conditions. Following the Great Fire of London in 1666, brick and stone became compulsory building materials, hence their wide use until the 1940s. Nevertheless due to post-war shortages of raw materials and labour, the demand for quick reconstruction and the advent of modernist architecture, construction turned to prefabricated building component solutions. Brick has recently been revived and this, according to Samuel, comes as a direct result of developers' aspirations to create a new housing class within city centres. This process of gentrification aims at converting warehouses into luxurious properties and preserving period facades, where in most cases brick is the main material. Therefore the needs of this market legitimise the return of brick as building material, for it reflects notions of pastness. Through this investment, the warehouse becomes a cultural artefact and its image as an urban ex-industrial space is entirely lost, submitting to the 'rules of the "picturesque"'.[69]

The ability of building materials to carry memories from the past is a major tool in the hands of architects and developers to manipulate public opinion and inspire sympathy. In the United Kingdom, the transition from the enormous, high-rise concrete social housing of the 1950s and 1960s to the small low-rise brick social housing of the 1970s onwards was partly legitimised by the return to brick as a building material. Bricks refer to earlier construction traditions that are greatly appreciated by the majority of the British public. In addition, there is an idea that the 'brick is the most "English" of the building materials: honest and down-to-earth, plain and unvarnished, it reflects the national virtues'.[70]

Considering the capacity of materials to withhold memory and convey meaning, it is surprising how the majority of recent memorials have been built of concrete, whilst concrete itself is most commonly associated with the erasure of memory.[71] In 'Concrete and Memory', Adrian Forty remembers Henri Lefebvre's views on concrete in relation to the historical continuity of cities: 'Here I cannot read the centuries, not time, nor the past, nor what is possible.'[72] Since the advent of modernist architecture, concrete has indeed been associated with oblivion, as modernism itself aimed at a break with the past. For Forty however this is not true. He argues that concrete is used, apart from obvious financial and structural reasons, because it is associated with notions of radical change and improvement. To support this opinion, he refers to the concrete *Märzgefallenen* memorial in a cemetery at Weimar that was designed by the Bauhaus director Walter Gropius. This political memorial that was built in 1922 and reconstructed in 1946 is dedicated to the working class of the Weimar Republic and has a simple form, for it is perceived as a symbol of collective effort. The bond between different components found in concrete refers to the bond between individuals, which altogether worked towards a common target, the unity of the working classes.[73] Similarly, a memorial to victims of the Fosse Ardeatine massacre in Rome, made out of concrete mixed with local aggregate, looks like an enormous piece of stone; thus in this case concrete not only imitates nature but also reproduces it on a much bigger scale.[74] Therefore concrete has the capacity to relate to the past and convey meaning as much as it implies forgetfulness.

The gradual development of heating systems from the open fire of the hearth to the cockle stove and fireplace, and later their subsequent replacement by a network of boiler and radiators, has been a long process that is the result of observations and experimentations. The American sociologist Richard Sennett argues that such a technological evolution 'revolves around three key issues: metamorphosis, presence, and anthropomorphosis'.[75] For instance, the steam heating system of the mill in Dornoch developed only after Snodgrass had 'observed a mode of drying muslins by wrapping them round hollow metal cylinders, filled with steam, practised at the bleach-fields near Glasgow, [and] it occurred to him, that by means of a proper apparatus, steam might be applied to heat a cotton mill, or any other large manufactory'.[76] This is the process of metamorphosis, which can develop in three directions. Firstly, it can entail the evolution of a *type-form*, which is 'a change [that] occurs through the elaboration of its species'.[77] As soon as heating systems technology produced no more negative results, it moved in new directions. Equally crucial in the evolution of a type-form is failure. Failure does not necessarily imply something negative but rather an incentive for further improvement. Metamorphosis can also occur when *different elements are integrated for the creation of new ones*. For instance, the heating of greenhouses was initially achieved by introducing a technology used in a residential setting (the stove) combined with the existing technology of greenhouses (capturing heat). Later on, the technology of heating greenhouses with a basement stove and a network of pipes circulating hot water and distributing warmth was applied to the heating of houses, which resulted in the popular boiler-radiator heating system. This shift where one practice is applied to a different area of interest is what Sennett calls a *domain shift*.

Materials also carry the trace of their creators, what Sennett calls *presence*. We have seen above how in Roman architecture, for instance, every new city was founded around a stone altar, the *mundus*, which burnt a fire lit from the initial altar in Rome. This fire symbolised the beginning of the new city and referred back to the founder Rome. Fire was similarly used in modern history to relate back to a place of origin, for example during the Olympic Games of 1936 in Nazi Germany. Hitler's aim to relate classical Greece to the Aryan race was devised by Dr Carl Diem, who introduced the relay of the Olympic flame.[78] Runners, one by one, carried a lit torch from ancient Olympia through Athens all the way to Berlin.

Materials also feature anthropomorphic qualities. These qualities manifest themselves in language, in expressions that people identify with, like an *environmentally friendly* material or the *honest colour* of a rendered wall. Commenting on the tendency to attribute ethical human qualities to materials, Sennett suggests that the 'purpose is to heighten our consciousness of the materials themselves and in this way to think about their value'.[79]

Typologies and memory

The word typology derives from the Greek word *typos*, which means imprint. *Typos* is a concept with a long history that underlies human knowledge since ancient times. Plato for instance perceived it as an idea, whereas Aristotle thought about it as a shape. In architecture the term typology is used as in the sciences, and refers to the result of a process that is based first on observation, then classification of different types of individual units into groups and then comparison between them. This process is an attempt to gain knowledge of the world we live in, and it can be either a concrete or a conceptual approach, although these two approaches are often blurred.

During the eighteenth century, there was increasing interest amongst natural historians in identifying common characteristics shared by different animals and plants. This interest led to the gradual creation of a theory of archetypes. The German school of philosopher-scientists, with Goethe as a leading figure, conducted research on various plants, as they were fascinated by the fact that different types shared many similarities. Their aspiration was to discover a common pattern amongst all plants, an archetype, and if that was possible then to expand the research and establish a common pattern amongst all species in nature. Goethe's initial approach was to create a universal plant-type, the *Urpflanze,* from which all plants were derived.[80]

This method of looking for common characteristics amongst species and categorising them accordingly was later translated and applied to architecture.[81] The first person to use the term typology in architectural history was the French writer Quatremère de Quincy in 1825. For him, typology was understood as a blend of reason, use and custom, and by applying it to architecture one could connect buildings with nature, society and tradition. Before him, in 1819, Jean-Nicolas-Louis Durand gathered and grouped the plans of important historical buildings, such as theatres, stadia and markets, drew them in a common scale and presented them in his plates of *Recueil et Parallèle des Edifices*. He wanted to explain the laws of composition on the basis of abstract models and at the same time to compare buildings that had the same use throughout different periods and areas. This way of classification was reminiscent of Goethe's work on plants, according to which types were generated by an archetype. Durand's intention was educational and, as Philip Steadman writes, his students

> by studying types and their underlying principles of composition, [should] be enabled to cope with what would otherwise be a bewildering and endless variety of individual, unrelated problems … if the student were to devote himself to mastering general type-forms, then he would be in a position to modify these types appropriately, to respond to the particular exigencies of function, site, budget and the requirements of individual clients.[82]

The fundamental difference between Quatremère's and Durand's understanding of typology is a matter of context. The former believed that a typology should be considered in context with relevant historical and cultural conditions, whereas the latter claimed that a typology is applied with no reference to original context. Nevertheless the way architecture was practised in the nineteenth century was according to Durand's prescriptions. Models and examples of architectural objects were classified according to their function and were to be found in manuals and handbooks, which architects followed faithfully.

The concept of architectural typology was reassessed in the 1960s after the publication of the art historian Giulio Carlo Argan's essay 'On the Typology of Architecture'. Until then architecture had treated typologies as compiled entries in manuals and catalogues, but the discontinuity with history propagated by the modern movement made their use redundant. Architects perceived them as obstacles to creativity, for they referred to previous architectural establishments. Instead architecture aspired to be mass-produced based on the repetition of a contemporary ideal type (unit). As Rafael Moneo comments, 'type had become prototype'.[83] The prototype of the modern movement, the unit to be blindly repeated, was quite different from how typology had been perceived in the past. Whereas Durand's type allowed a degree of flexibility and adjustability depending on site and use, the modern prototype excluded even that.

Because of this Argan tried to re-establish the concept of typology and to return to Quatremère's theory. He supported the idea that a typology is not an image to be copied and imitated, as this is the role of a model. Instead typology is vague, for it is an abstraction. It demands the elimination of every architectural element that is not common amongst the buildings in comparison. Therefore when applied to architecture, it does not affect it in the same way as the blind replication of a model. As a mere reduction of the original, typology is also a mere reduction of history itself. In this light, although typologies carry residues of the past, they are prey to the architect's interpretation and ability to adjust them to the contemporary context. Defending his views on typology and the way it should be adapted that respects the spirit of modernism, Argan concludes that '[t]ypological and the inventive aspects of the creative process are continuous and interlaced – the inventive aspect being merely that of dealing with the demands of the actual historical situation by criticising and overcoming past solutions deposited and synthesized schematically in the "type"'.[84]

In 1966 the Italian architect Aldo Rossi developed a different approach to the concept of typology in *The Architecture of the City*. For Rossi, as types are to be found in all architectural creations they should be treated as cultural objects. In this sense, studying types is a way to analyse architecture. He perceives types as having a predetermined form but they do not manifest themselves in isolation, but rather in a constant dialogue with techniques, functions and styles. Therefore there are *only* variations of initial typologies that are adjusted to contemporary settings. In effect, Rossi claims, '[t]ype is the very idea of architecture, that which is closest to its essence. In spite of changes, it has always imposed itself on the "feeling and reason" as the principle of architecture and of the city.'[85]

In his essay 'Typology and Design Method', Alan Colquhoun presents a different view on typology. Typology in architecture, Colquhoun argues, is necessary, for it is a means of communication. He makes the assumption that if a building makes no references to types from the past then its form is a new language of communication, invented arbitrarily by the architect. Therefore an architect has the ability to convey meaning and ideology through design, regardless of whether reference is made to types or not.[86]

A short account of the history of typology reveals a glimpse of the confusion in the discipline of architectural history and theory with regard to its definition. In any case all theories of typologies share a common characteristic. Each typology reflects some sort of memory that is present in both buildings and people's minds. As John Ruskin eloquently writes about architecture, '[w]e may live without her, and worship without her, but we cannot remember without her'.[87] I therefore argue that a typology depends upon the way one looks at architecture. If a building is perceived and classified as an autonomous, self-sustained *unique* object, this does not allow for any further reductions. It is classified through its own uniqueness. On the other hand, if a building is perceived as an ordered assembly of various smaller objects, these objects are not unique, as they can be identified in other buildings too. In other words, architectural objects are there to be repeated. In this case if one wants to recall a particular object, like a roof, in the process of recollection he brings to mind a series of various types of roof that he has previously seen. The way memory operates therefore is by typifying objects. This follows the functioning of language: the process of naming an object as a roof implies that this object shares common characteristics with other roofs, which comes to be confirmed through memory, and thus the nature of naming entails a typifying process in its very essence. Typologies allow for the possibility of grouping different objects together. While they appear to reduce the memory of the whole building, they simultaneously encompass an entire group of other objects that have the same characteristics. Typologies therefore function as archives of either reduced or enhanced memories and through repetition they carry these memories further into the future.

Architectural evolution and fire: A Bachelardian approach

I have so far outlined key stages of architectural evolution based on the way architecture has embraced the element of fire since the time of the primitive hut. Supported by an energetic reading of architectural evolution, this study has further elaborated on architecture's ability to record memory. Architecture appears to function as a porous entity that absorbs the needs and demands of its users. Architectural configurations, technologies, materials and typologies are judged according to their efficiency and success. Through observation of their performance one acquires knowledge that can be used to inform future advancements. Then, through processes of experimentation, new arrangements or techniques, which already carry the memory of previous adaptations, are tried out to either improve or replace existing establishments. Architecture therefore operates as an open-ended, constantly updating archive that reflects the spirit of its era.

This archival reading of architecture agrees with Bachelard, who questions: 'What remains of the historical past, what lasts from it? Only that which has reasons for beginning again.'[88] This study breaks down the duration of architectural evolution to instants of creation. Out of nothingness, architecture evolves ceaselessly in time, with alternating periods of affirmation and negation. Evolution depends on the rhythm of these instants created by periods of work and repose. It is

> driven by the will to perpetuate or 'copy' ourselves. However, each time we set ourselves to this task, we must exceed or improve on this 'copy', or else it will dissolve itself ... The moment of remaking breaks the flow of time – repetition, to survive, must involve a negation, as well as an affirmation, of previous repetitions. Rhythm is not an 'eternal return', but an 'eternal reprise'.[89]

Whatever appears to be efficient and functional will be repeated in future architectural creations, whereas failure is more likely to be subject to further improvement or replacement. The memory that the architectural archive stores is useful knowledge. Following Bachelard's views on knowledge production, knowledge advances dialectically by overcoming obstacles of existing and past knowledge. Architecture emerges and evolves out of a ceaseless *no* to these obstacles because in general 'everything comes alive when contradictions accumulate'.[90]

After a critical analysis of buildings, new experimentation takes place, and this opens up a dialogue between buildings and people, and amongst people themselves, which agrees with Bachelard's view that knowledge must be social to advance. At the same time the fact that it is a product of dialogue also indicates that this knowledge has an open-ended character, thus it can never stay static. Lastly, as this knowledge reflects the memory of architectural evolution and the latter has been broken down to instants of creation, this implies further that knowledge is historically located. For Bachelard, the fact that knowledge is historical lends it a dynamic character as 'it advances through a tension between present and future, or between conservation and change ... [knowledge] has an aim, "it has a destiny."'[91] Through a Bachelardian approach to architectural evolution, architecture operates as an archive in which a reduced memory of its entire past is stored and recorded. This memory functions as a guide for future architectural creations and is constantly updated every time a new building is built.

Architectural evolution and fire: A Bergsonian approach

Bergson's theory of evolution is diametrically opposed to that of Bachelard. The latter's approach to evolution reflects a conventional understanding in Western thinking according to which entities are experienced and studied in isolation, statically, and in chronological order. Bergson's theory is a reaction to the mechanism of neo-Darwinism and the finalism of neo-Lamarckism.[92] Both mechanism and finalism perceive and evaluate evolution in the present and imply a constant invention of a product that must always be new. Bergson, however, perceives evolution in the same way as Nietzsche. Evolution for him is not merely a linear, mechanistic process during which an organ, for instance, responds and adapts itself to external conditions, but a dynamic process in flux. Every evolution is driven by *élan vital*, the vital force or impulse that keeps this process constantly changing. The process is not finite, for it has no end and no end product. In fact, Bergson's evolution, being clearly influenced by Lucretius' philosophy,[93] is a creative, unpredictable, open-ended evolution with its own becoming or, as the philosopher Keith Ansell-Pearson describes, a creative evolution: 'a play between process and product'.[94] As everything unfolds in time, the same evolution can never take place twice. In this sense the repetition of a process is impossible unless we refer to it in abstract terms.

The understanding of Bergson's creative evolution presupposes a clear distinction between the two types of time in his philosophy. As previously discussed, there is a homogeneous time, which is the fragmented time of physics, and a heterogeneous one, which is real time, or duration. The time of physics, which is the time Bachelard relies on, is spatialised time. It is projected in space so that it can obtain a number, it can be measured. In *Time and Free Will* Bergson discusses how numbers imply discontinuity. Every number is a single unit. The number 3, for instance, is a unit because it has a name, which is 3 or three. At the same, this number can also be considered as a multiplicity of other units that are identical to each other (1+1+1). The fact that we can count these units and add up to the number 3 means that we no longer perceive them as single units, as this would lead to their merging into a single unit whilst still being distinct from each other. In order to count these units we juxtapose them in homogeneous space. Similarly, the time of physics is counted in the way that a number is counted. It is projected in space and split into an invented succession of homogeneous instants, all of which added up give the measurement of time. Real time or duration is very different from the time of physics and can only be perceived with the method of intuition.

> [D]uration is not merely one instant replacing each other; if it were, there would never be anything but the present – no prolonging of the past into the actual, no evolution, no concrete duration. Duration is the continuous progress of the past which gnaws into the future and which swells as it advances.[95]

In *Creative Evolution*, Bergson explains that our consciousness consists of two interdependent tendencies: instinct and intelligence. Our inability to perceive the flow of duration, and consequently the duration of evolution, lies in a problem of our intelligence. Whereas duration can be experienced instinctively, we tend to perceive it with our intellect; thus we break it down to fragments, instants, as if it were the time of physics. Nevertheless these instants are merely snapshots from the continuity of duration that the intellect has extracted from it.

Instinct and intelligence are one of Bergson's dualisms. The distinction between them can help unpack his ideas about evolution. Commenting on the evolution of plants and animals, Bergson claims three tendencies of the *élan vital* as common to both plants animals – vegetative torpor, instinct and intelligence. In fact, he writes that

> [t]he cardinal error, which from Aristotle onwards, has vitiated most of the philosophies of nature, is to see in vegetative, instinctive and rational life, three successive degrees of the development of one and the same tendency, whereas they are three divergent directions of an activity that has split up as it grew. The difference between them is not a difference of intensity, nor, more generally, of degree, but of kind.[96]

Vegetable and animal life are interdependent and their tendencies merge. This means that in plants one can detect traces of instinct and intelligence in the form of mobility and consciousness that can be awakened, whereas in animals the potential of reaching a vegetative state is always a possibility. Likewise intelligence and instinct, which are tendencies amongst animals, interpenetrate each other and share characteristics of their common origin. The only difference is the proportions of each tendency from animal to animal. Bergson illustrates intelligence and instinct using examples from biology and gives a clear definition: 'Instinct perfected is a faculty of using and even of constructing organised instruments; intelligence perfected is the faculty of making and using unorganised instruments.'[97] Both instinct and intelligence possess innate knowledge. The difference is that the former carries innate knowledge of things, of matter, and the latter of relations,[98] which is the knowledge of how things relate to each other, of their form. At this point, Bergson questions whether it is possible to possess knowledge that has form but no matter. His answer confirms it, and in fact he posits that this is the advantage of the formal knowledge of intelligence compared to the material knowledge of instinct. Bergson explains that

> form, just because it is empty, may be filled at will with any number of things in turn, even with those that are of no use. So that a formal knowledge is not limited to what is practically useful, although it is in view of practical utility that it has made its appearance in the world. An intelligent being bears within himself the means to transcend his own nature.[99]

The intellect has a tendency to construct. Every fabrication it renders is of inert matter. Even if it fabricates organic, living matter, the intellect will treat it as if it was inert. In other words, 'fabrication deals only with the solid; the rest escapes by its very fluidity'.[100] The fact that our intellect deals with inert matter has an advantage. This is the freedom to divide this matter into as many parts as one pleases, thus making it discontinuous. Nevertheless this freedom of unlimited discontinuity that one can impose on matter gives the impression of continuity. The reason intellect does this is because '[o]f the discontinuous alone does the intellect form a clear idea'.[101] Therefore it is a tendency of the intellect to perceive both fluidity and the duration of evolution as discontinuous. Only if movement is broken down into immobilities is the intellect able to understand it. This function of the intellect is for Bergson cinematographic.

Applying Bergson's evolutionary thinking to architecture means that the totality of architecture's past exists in the present, constantly pointing towards the future. Past and present coexist and each present returns back to itself as past. Architecture, when perceived in duration, in flux, exists as a *virtual* multiplicity. It becomes an *actual* one only in the present because of the necessity of our intellect to break down its continuity into immobilities to understand it. To illustrate the difference between virtual and actual multiplicities, Bergson in *Time and Free Will* refers to the colours of the rainbow.[102] Commenting on Bergson's rainbow, the archaeologist Johan Normark writes:

> The rainbow has a multiplicity of heterogeneous colours. There is a continuum of differentiated change in its spectrum. This characterises a virtual multiplicity in time. The colours cannot be individuated since they constantly move into another colour. If we do define the colours, we separate them from each other, give them boundaries and a fixed spatial location. Then they become numerical, metric, homogenous and actual. These can be counted and analysed according to our needs, but they are misrepresenting the true heterogeneous character of the colours of the rainbow.[103]

If one examines the duration of architectural evolution purely from a Bergsonian perspective then architecture must be understood as a virtual multiplicity, for only then the totality of its past is preserved in the present. Otherwise, if architecture is perceived as a group of autonomous objects, of actual multiplicities like the colours of the rainbow individually, devoid of duration, its evolution is broken down into immobilities and its virtual past is erased. In other words, a Bergsonian understanding of architectural evolution confirms that a building perceived at any moment of its duration functions as an archive that withholds the totality of its entire past.

Notes

1. In an essay entitled 'Primitive, the Word and the Concept', Adrian Forty questions the concept of 'primitive' in architecture. He explains that this concept is different from how it is understood in painting, as the latter never relied on its origins to justify itself. Forty points out that since the time of the Roman architect Vitruvius, architectural theory has referred to the word 'primitive' but it did not then have the same meaning as today. Equally, Semper's understanding of the primitive hut, which is the Caribbean hut, does not conform to our contemporary understanding of the word. In all cases, 'primitive' referred to the word 'original' until the beginning of the twentieth century. Then the word obtained a new meaning; Forty uses as an example Adolf Loos's essay 'Ornament and Crime'. According to Forty, '[f]rom Loos's time onwards, the word "primitive" has been taken up positively as a model to follow or negatively.' Adrian Forty, 'Primitive, the Word and the Concept' in Jo Odgers et al., eds, *Primitive: Original Matters in Architecture* (New York: Routledge, 2006), 6.
2. Thomas Gordon Smith, ed., *Vitruvius on Architecture* (New York: Monacelli Press, 2003), 38.
3. Gottfried Semper, *The Four Elements of Architecture and Other Writings* (Cambridge and New York: Cambridge University Press, 1989), 102.
4. Semper, *The Four Elements of Architecture and Other Writings*, 102.

5. The mundus is often associated with *Umbilicus Urbis Romae* (Navel of the City of Rome).
6. Joseph Rykwert, *The Idea of a Town: The Anthropology of Urban Form in Rome, Italy and the Ancient World* (London: Faber and Faber, 1976), 104–5.
7. F. G. Asenjo, *In-Between: An Essay on Categories* (Washington, DC: Center for Advanced Research in Phenomenology and University Press of America, 1988), 11.
8. Asenjo, *In-Between*, 10.
9. Lisa Heschong, *Thermal Delight in Architecture* (Cambridge, MA: MIT Press, 1979), 29.
10. Heschong, *Thermal Delight in Architecture*, 36.
11. Philippe Rahm Architectes, 'Digestible Gulf Stream', http://www.philipperahm.com/data/projects/digestiblegulfstream/index.html, accessed 1 March 2018.
12. Philippe Rahm Architectes, 'Digestible Gulf Stream'.
13. L. Wright, *Home Fires Burning: The History of Domestic Heating and Cooking* (London: Routledge and Kegan Paul, 1964), 6.
14. Wright, *Home Fires Burning*, 28.
15. John Alfred Gotch, *The Growth of the English House: A Short History of its Architectural Development from 1100 to 1800* (London: Batsford, 1909), 106.
16. Wright, *Home Fires Burning*, 19.
17. Wright, *Home Fires Burning*, 19–20.
18. Johann Beckmann, *A History of Inventions and Discoveries*, vol. 2, translated by William Johnston (London, 1817), 103.
19. John Lyons, *Fire* (New York: Scientific American Books, 1985), 47.
20. Wright, *Home Fires Burning*, 21.
21. Wright, *Home Fires Burning*, 58.
22. Stephen J. Pyne, *Fire: A Brief History* (London: British Museum Press, 2001), 124.
23. The name of phosphorus comes from the Greek words *phos* meaning light and *phoros* meaning bringing.
24. John Emsley, *The Shocking History of Phosphorus: A Biography of the Devil's Element* (London: Macmillan, 2000), 4–5.
25. Emsley, *The Shocking History of Phosphorus*, 67.
26. Jaime Wisniak, 'Matches: The Manufacture of Fire', *Indian Journal of Chemical Technology*, 12 (May 2005), 372.
27. Emsley, *The Shocking History of Phosphorus*, 69–70.
28. Wisniak, 'Matches: The Manufacture of Fire', 372.
29. Wisniak, 'Matches: The Manufacture of Fire', 375–6.
30. Emsley, *The Shocking History of Phosphorus*, 77–8.
31. R. Bean, B. W. Olesen and K. W. Kim, 'History of Radiant Heating and Cooling Systems Part 1', *ASHRAE Journal*, 52: 1 (2010), 40.
32. Qinghua Guo, *Chinese Architecture and Planning: Ideas, Methods, Techniques* (Stuttgart and London: Edition Axel Menges, 2005), 19–27.
33. Bean, Olesen and Kim, 'History of Radiant Heating and Cooling Systems Part 1', 41–2.
34. Lyons, *Fire*, 47.
35. Robert Bruegmann, 'Central Heating and Forced Ventilation: Origins and Effects on Architectural Design', *Journal of the Society of Architectural Historians*, 37: 3 (October 1978), 144.
36. Louis Fernández-Galiano, *Fire and Memory: On Architecture and Memory*, translated by Gina Cariño (Cambridge, MA: MIT Press, 2000), 224.
37. Bruegmann, 'Central Heating and Forced Ventilation', 144.
38. Bruegmann, 'Central Heating and Forced Ventilation', 146.
39. Bruegmann, 'Central Heating and Forced Ventilation', 147.
40. *Transactions of the Society, Instituted at London, for the Encouragement of Arts, Manufactures, and Commerce*, 24 (1806), 118–19.
41. Bruegmann, 'Central Heating and Forced Ventilation', 147.
42. Emmanuelle Gallo, 'Jean Simon Bonnemain (1743–1830) and the Origins of Hot Water Central Heating', in Malcolm Dunkeld et al., eds, *Proceedings of the Second International Congress on Construction History* (Cambridge: Construction History Society, 2006), 1043.
43. Bruegmann, 'Central Heating and Forced Ventilation', 148.
44. Watson and Pritchett Architects, *Plans, Elevations, Sections, and Description of the Pauper Lunatic Asylum … at Wakefield, … To which is Added a New and Enlarged Edition of S. Tuke's Practical Hints on the Construction and Economy of Pauper Lunatic Asylums* (York, 1819).

45. Surveyor-General of Prisons, 'Report of the Surveyor-General of Prisons on the Construction, Ventilation, and Details of Pentonville Prison' (London, 1844).
46. Fernández-Galiano, *Fire and Memory*, 229.
47. Katherine Shonfield, *Walls Have Feelings: Architecture, Film, and the City* (London and New York: Routledge, 2000), 4.
48. Helen Molesworth, 'Bathrooms and Kitchens: Cleaning House with Duchamp' in Nadir Lahiji and Daniel S Friedman, eds, *Plumbing: Sounding Modern Architecture* (New York: Princeton Architectural Press, 1997), 76–7.
49. Molesworth, 'Bathrooms and Kitchens', 77.
50. Yan Ji and Stelios Plainiotis, eds, *Design for Sustainability* (Beijing: China Architecture and Building Press, 2006), 2–3.
51. T. J. Williamson, Antony Radford and Helen Bennetts, *Understanding Sustainable Architecture* (New York: Spon Press, 2003), 1.
52. Ken Yeang, 'A Theory of Ecological Design' in William W Braham, Jonathan A Hale and John Stanislav Sadar, eds, *Rethinking Technology: A Reader in Architectural Theory* (London and New York: Routledge, 2007), 391.
53. Ji and Plainiotis, *Design for Sustainability*, 5.
54. Xenophon, *Memorabilia*, translated by Amy L. Bonnett (Ithaca, NY and London: Cornell University Press, 2001), 93.
55. Dominique Fretin and Gilda Collet Bruna, 'Solar Energy: Harvesting the Sun. An Architectural Approach Through Le Corbusier's Groundings', *Engevista*, 13: 1 (May 2011), 5.
56. Ji and Plainiotis, *Design for Sustainability*, 11–13.
57. Ji and Plainiotis, *Design for Sustainability*, 15–16.
58. Ronald DiPippo, *Geothermal Power Plants: Principles, Applications and Case Studies* (Oxford and Heidelberg: Elsevier, 2009), 20–1.
59. DiPippo, *Geothermal Power Plants*, 61.
60. Ramon Margalef, *Perspectives in Ecological Theory* (Chicago and London: University of Chicago Press, 1968), 2.
61. Margalef, *Perspectives in Ecological Theory*, 2–3.
62. Fernández-Galiano, *Fire and Memory*, 229.
63. Neil Leach, ed., *Rethinking Architecture: A Reader in Cultural Theory* (New York: Routledge, 1997), 20.
64. Michael Richardson, *Georges Bataille* (London and New York: Routledge, 1994), 123–4.
65. Georges Bataille, *The Accursed Share*, vol. 1 (New York: Zone, 2002), 21.
66. Bataille, *The Accursed Share*, 30.
67. Pyne, *Fire*, 18.
68. Raphael Samuel, *Past and Present in Contemporary Culture* (London: Verso, 1999), 128.
69. Sharon Zukin, *Loft Living: Culture and Capital in Urban Change* (Baltimore and London: Johns Hopkins University Press, 1982), 180.
70. Samuel, *Past and Present in Contemporary Culture*, 129.
71. Adrian Forty, 'Concrete and Memory' in Mark Crinson, ed., *Urban Memory: History and Amnesia in the Modern City* (London and New York: Routledge, 2005), 78.
72. Henri Lefebvre, *Introduction to Modernity*, translated by J. Moore (London and New York: Verso, 1995), 119.
73. Forty, 'Concrete and Memory', 86.
74. Forty, 'Concrete and Memory', 89.
75. Richard Sennett, *The Craftsman* (London: Penguin, 2009), 120.
76. *Transactions of the Society, Instituted at London, for the Encouragement of Arts, Manufactures, and Commerce*, vol. 24 (1806), 118–19.
77. *Transactions of the Society, Instituted at London, for the Encouragement of Arts, Manufactures, and Commerce*, vol. 24 (1806), 125.
78. Nigel Jonathan Spivey, *The Ancient Olympics: War Minus the Shooting* (Oxford: Oxford University Press, 2004), 247.
79. Sennett, *The Craftsman*, 137.
80. Philip Steadman, *The Evolution of Designs: Biological Analogy in Architecture and Applied Arts* (Cambridge: Cambridge University Press, 1979), 23–4.
81. As the term typology in architecture is used in the same way as in the sciences, one can develop a theory of architectural evolution through a scientific approach and specifically through a biological perspective. According to the American scientist Alfred Lotka (1880–1949), human

evolution manifests itself through two channels. On one hand, there are processes based on genetic material, which are called *endosomatic*. They occur within the body and they are based on the Darwinist theory of natural selection. On the other hand, there are other processes that are external to the body, the *exosomatic* ones, and they are understood in a Lamarckian sense. Nevertheless, as Philip Steadman argues, human evolutionary theory can also be applied to artefacts – thus also in architecture – and '[t]his is in the proposal that tools, or machines, or other implements, are in fact part of the evolution of one particular species of animal, i.e. man; that they are, so to speak, extensions of or substitutes for the various organs of his own physical body' (Steadman, *The Evolution of Designs*, 124). A biological interpretation of an *endosomatic* process in architecture, or else a process that occurs *within* architecture itself, refers to the typology of a building, which, as I previously mentioned, functions as memory that is present in buildings and in people's minds.

82. Steadman, *The Evolution of Designs*, 29–31.
83. José Rafael Moneo, *Theoretical Anxiety and Design Strategies in the Work of Eight Contemporary Architects* (Cambridge, MA: MIT Press, 2004), 33.
84. G. C. Argan, 'On the Typology of Architecture', translated by J. Rykwert, in *Theorizing a New Agenda for Architecture: An Anthology of Architectural Theory, 1965–1995*, ed. Kate Nesbit (New York: Princeton Architectural Press, 1996), 246.
85. Aldo Rossi, *The Architecture of the City* (Cambridge, MA and London: MIT Press, 1982), 41.
86. Alan Colquhoun, 'Typology and Design Method', *Perspecta*, 12 (1969), 71–4.
87. John Ruskin, *The Seven Lamps of Architecture* (Elibron Classics, 2005 [1849]), 324.
88. Gaston Bachelard, *The Dialectic of Duration*, translated by Mary McAllester (Manchester: Clinamen, 2000), 20.
89. Conrad Russell, 'Fictive Time: Bachelard on Memory, Duration and Consciousness', *Kronoscope*, 5: 1 (2005), 7.
90. Gaston Bachelard, *The Poetics of Space*, translated by Maria Jolas, with foreword by John R. Stilgoe (Boston: Beacon Press, 1994), 39.
91. Cristina Chimisso, *Gaston Bachelard, Critic of Science and the Imagination* (London and New York: Routledge, 2001), 124.
92. In *Creative Evolution*, Bergson discusses the evolution of the eye as an organ. A mechanistic evolutionary theory perceives that this organ has evolved by choosing the best components to construct itself, whereas a theory of finality perceives that this organ is the product of a pre-conceived plan that aimed at a specific end.
93. Lucretius (first century BC) made a significant contribution to philosophy, at least as it is understood today, because of his early understanding of an evolutionary theory. In his poem *De Rerum Natura* (*On the Nature of Things*) Lucretius explains the universe through our perception of the invisible movement of atoms. Bergson, in his book *The Philosophy of Poetry: The Genius of Lucretius*, points out Lucretius' evolutionary understanding of the universe, according to which there is a vital continuity at play, reminiscent of Bergson's duration.
94. John Mullarkey, ed., *The New Bergson* (Manchester and New York: Manchester University Press, 1999), 149.
95. Henri Bergson, *Creative Evolution*, translated by A. Mitchell (New York: Henry Holt, 1911), 11.
96. Bergson, *Creative Evolution*, 79.
97. Bergson, *Creative Evolution*, 82.
98. Bergson, *Creative Evolution*, 86.
99. Bergson, *Creative Evolution*, 87.
100. Bergson, *Creative Evolution*, 88.
101. Bergson, *Creative Evolution*, 89.
102. Henri Bergson, *Time and Free Will* (London: Elibron Classics, 2005), 66.
103. Johan Normark, 'Time and Archaeology 4: Bergson's Continuous Duration', http://posthumanocentric.webs.com/apps/blog/show/604204–time–and–archaeology–4–bergson–s–continuous–duration, accessed 25 May 2018.

6
Architecture on Fire

> For the speedy restoration whereof, and for the better regulation, uniformity and gracefulness of such new buildings as shall be erected for habitations in order thereunto; and to the end that great and outrageous fires (through the blessing of almighty God) so far forth as human providence (with submission to the divine pleasure) can foresee, may be reasonably prevented or obviated for the time to come, both by the matter and form of such building.
>
> Rebuilding of London Act 1666 (18 & 19 C. II. c.7.)

Fire is an ongoing threat to architecture. Although buildings and societies are often well prepared to combat fire, catastrophes still occur. A spark might be the cause of the complete collapse of a building, a neighbourhood or, as seen throughout history, of entire cities.[1] The moment when a fire breaks out in a building signifies the beginning of a process. The process of combustion demands the simultaneous working of three ingredients in order to occur.

This (fig. 6.1) is fire's eternal triangle: the three corners are occupied by oxygen, a combustible material and a certain level of temperature. In a balanced system of any two of the elements, the introduction of the third leads to the beginning of a combustion process. Fire's eternal triangle is formed as soon as a specific heat level is achieved. For the flaming process to stop, one of the triangle's components needs to be removed.[2] Recently, following the introduction of chemical fire extinguishing, the fire triangle is no longer representable, as it excludes the chain reaction of chemicals. A new diagram has been created to accommodate all four aspects of fire's combustion and extinguishing.

This is the *tetrahedron of fire*, which establishes combustion as a chemical process (fig. 6.2). Each triangular surface of the pyramid represents a necessary element for combustion to take place. As soon as one of these elements is removed, the combustion process finishes.[3]

One of the most common practices for extinguishing fire is throwing water on flames. This helps to reduce the local temperature below that at which combustion is possible. Equally, deposits of water thrown around the periphery of a flaming centre will prevent its expansion. The

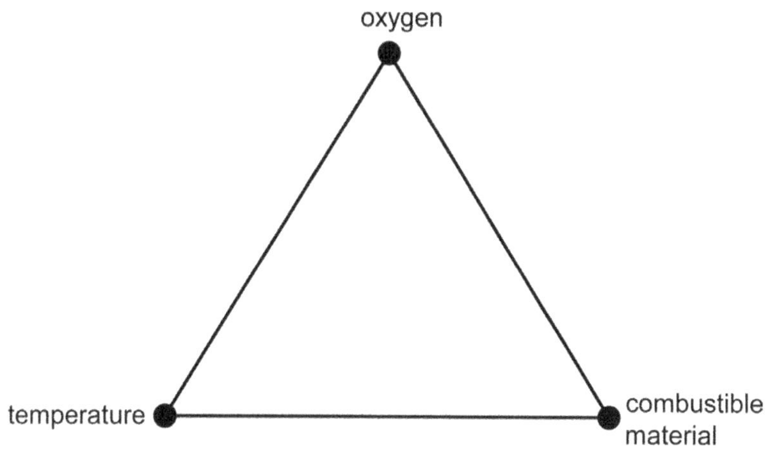

Figure 6.1 Fire's eternal triangle. © Stamatis Zografos.

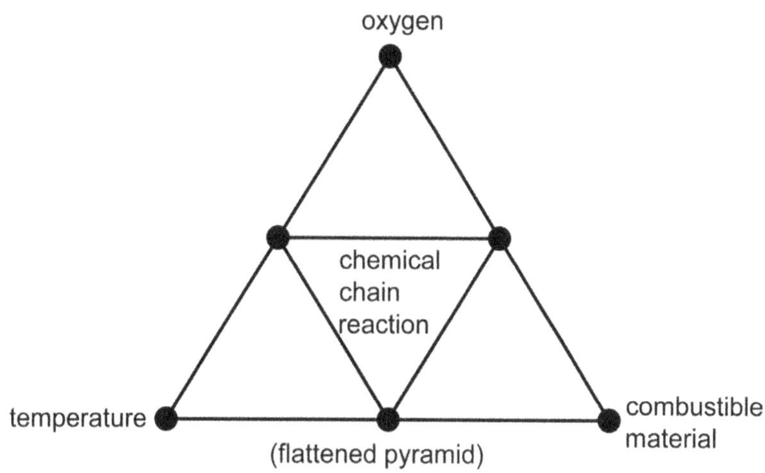

Figure 6.2 The tetrahedron of fire. © Stamatis Zografos.

introduction of water to extinguish fire may also have an adverse effect. Water on a deep oil fire will suffocate the fire temporarily but the oil, as it reaches a high temperature, will scatter and reignite it. A different process of extinguishing fire is either cooling the gaseous combustion area or removing the combustible itself from its environment. Alternatively, one can also introduce a barrier between the burning surface and the flame, as in the case of a fireproof blanket, or by cutting off the air supply so oxygen does not feed the fire further.[4]

A normal process of combustion develops through certain stages, with the moment of ignition indicating the beginning. The first evidence of combustion is the smouldering stage. This is a non-flaming process accompanied occasionally by incandescence and in most cases by the emission of smoke. The next stage is the appearance of, predominantly, smoke, or else the emission of very hot fine particles indicating an incomplete combustion, which may only reach temperatures of 80–100°C, but can still be lethal. Next is the stage of incandescence, a glowing heat. The final stage of a developing combustion is flaming, which refers to the oxidation of liberated gases generating heat and light. It indicates the maximum growth of a combustion process before its decay begins.[5] Flames produce a temperature of the order of 1500K (*c.* 1200°C) and maximum of 2500K (*c.* 2200°C) in air.[6]

Within a building, the rise of temperature instigated by human intervention or an electrical overloading may lead to a tragic event. If the level of heat reaches the smouldering stage, the fuel material starts decomposing. This is the stage of an early combustion, which, although it does not involve flaming of the material, in most cases releases potentially lethal smoke. The smouldering combustion can spread to adjacent materials reaching temperatures of up to 400°C (which is usually the temperature of a flaming combustion) and develop to full combustion.[7] In principle, the development of a combustion process depends upon the amount of oxygen present in space. Therefore a fire in a large open-plan and well-ventilated space develops rapidly and exhausts itself quickly, whereas in a building composed of a series of small and confined spaces where oxygen is limited, such as ducts, cupboards, ceiling voids etc., combustion is partial and the biggest impact is caused by the lethal toxic release of smoke.[8] In the latter scenario – the build-up of smoke in a confined space – as soon as it is noticed and a door is opened, the sudden wave of air causes an explosive flashover, which is the 'stage of development of a contained fire at which fire spreads rapidly to give large merged flames throughout the space'.[9] When flashover happens, every object and surface erupts into flames and the pressure produced might reach such an intensity that it is capable of blowing out windows. The expansion of fire in a building may initially develop steadily and in a predictable way but flashover is one of the most troubling aspects of fire's expansion in space. Whereas people are adjusted to linear phenomena and would expect a steady and predictable development, the sudden and explosive nature of flashover leaves fire witnesses entirely staggered.[10]

Statistically the most common reason for a fire to spread in a building is the building's contents, such as the furniture, fittings and fabrics, which all carry rapid and intense burning characteristics. Carpets and curtains in particular promote an unusually rapid spread throughout a building due to their proximity to the structure and its external skin.[11] The materials used for the construction of a building can also be a potential fire threat, and their choice must therefore abide by relevant building legislation. The resistance of materials to ignition is dependent upon their chemical composition and the surface area that is exposed to air. Modern man-made materials, in comparison with natural ones, tend to have a higher tolerance to flammability. Of all man-made materials, however, plastic, which is extensively used in contemporary buildings, has a peculiar reaction to combustion and requires particular attention as to how and where it is used. For instance, polyvinyl chloride (PVC), which is used to wrap electric cables (fig. 6.3), can cause fire if electrical overloading causes the plastic's thermal decomposition.[12]

PVC drainage pipework can similarly help fire spread between the spaces it serves, as it provides passage for smoke and a thoroughfare for heat and flames to travel through.[13] Regarding the external fabric of buildings, the heat and fire tolerance of cladding materials plays a

Figure 6.3 The affected electrical board and the overheated wiring that caused the fire in a branch of Barclays Bank in Gracechurch Street, City of London. © Mary Evans Picture Library.

significant role in the event of a fire, as they are the main means through which flames can spread to adjoining spaces. Ironically, the external skin of low-rise buildings is traditionally made of brick or stone, which have high heat tolerance, whereas contemporary high-rise constructions make use of low heat-tolerance cladding materials, which can help fire spread rapidly, creating irreversible damage.

Fire can also cause the complete collapse of a building if heat affects the structural frame. This can occur even when combustion has not reached its highest point but has been sufficient to distort the structure due to thermal expansion (fig. 6.4). The steel reinforcement of walls and slabs may expand to such an extent that bricks and concrete are demolished, bringing about ultimate collapse.[14]

When fire breaks out, it needs space to develop. The process of combustion produces hot smoke and gases that lead to the build-up of higher atmospheric pressure than the ambient one. Smoke travels towards zones of lower pressure and contaminates them. Heat and incomplete products of combustion in the smoke are the most frequent cause of fire spread. In large spaces, a micro-climatic condition is created when smoke and gases create intense air circulation to any space with low pressure.[15] In principle, any building is a potential space where fire can spread both horizontally and vertically. A vertical circulation, such as in a staircase or lift shaft, creates a flue effect that allows an easy expansion of fire. Equally,

Figure 6.4 Massive fire damage at Battersby hat factory in Offerton, Stockport, after a fire in 1906, showing the extent of destruction, yet manufacture was stopped for only six weeks. © Mary Evans Picture Library.

a corridor or a ceiling void can act as a horizontal flue, hence the broad use of fire-resisting doors and compartmentation. Amongst all horizontal spaces that may form a fire passage, voids over suspended ceilings are the most problematic. They usually conceal electrical and other services, which are high fire risks in themselves. Combustion can take place out of sight inside these spaces and can lead to the collapse of the ceiling before it is even noticed, or can create a flashover. These voids are spaces usually composed of and insulated with combustible materials, and they connect different parts of a building through service outlets, thus raising the risk of spreading flames throughout its entirety.[16]

Fire-fighting bodies

Since the Industrial Revolution, there has been an ever increasing concentration of combustible materials in every city. Different types of fuel are housed not only in tank farms and harbours but also in every corner of a city, every factory, shop, office and residential building. Fuel is also present outside buildings, in cars that are parked or moving about in front of them. The potential of fire accidents in cities is thus higher than ever before.

The threat imposed by fire is not a recent phenomenon. Since the early years of civilisation, when methods of kindling fire were invented and widely applied, fire accidents have been common. Prior to the invention of fire-fighting techniques, fire burnt until it died out, for the only means for extinguishing it involved water stored in each household, and that was never sufficient.[17] In response to the risk of fire, societies turned to the formation of groups to fight it. The earliest evidence of fire-fighting bodies is from Roman times.[18] The Roman fire-fighting force, which was composed of slaves, was called *Familia Publica*[19] and had stations by the walls and gates of the city.[20] Nevertheless the efficiency of the force to combat fire did not meet the Emperor Augustus' expectations and so in 6 AD he imposed fire-fighting reforms. The new force was called the Corps of Vigiles and was in action for 500 years. Its organisation and equipment are worth mentioning, as they were highly sophisticated. The total area of Rome was divided into fourteen regions. There were seven groups of *Vigiles* and each of them was responsible for two regions. Each group had an army of 1000 men that was split into ten smaller teams of 100 men. In other words, 7000 fire-fighters guarded a city of less than a million people, so there was one fireman per 140 citizens. Their stations were built strategically on the borderlines of two adjoining regions where faster approach to fire centres could be achieved.[21] An excavation that took place in 1820 revealed a sub-station that contained two statue bases with the inscriptions of firemen's different posts. Their duties were equal to those of contemporary fire-fighters. The *Aquarius* carried water, the *Siphonarius* supervised the pump and the *Uncienarius* lifted burning roofs with fire hooks.[22] Roman fire-fighters used large and heavy hammers to break doors, give access to burning buildings and even create fire breaks. They also used ladders to access the roof level and to take it down with a hook, and used axes to open locked doors.[23] Regardless of how primitive these tools seem, they have not changed and are still in use today.

The early Roman architect Vitruvius recorded the existence of a brass fire engine invented by a Greek inventor of the second century BC called Ctesibius. In the history of fire-fighting, Ctesibius' machine appears to be the first to make use of an enclosed air chamber as a means to provide a continuous flow of water.[24] Later, Heron of Alexandria, one of Ctesibius' pupils, invented a similar machine that included two vertical cylinders. The downward movement of these cylinders together with the assistance of a pump provided a continuous flow of water that discharged through a pipe fitted with a swivel joint so that water could be directed towards the fire. This pipe was called a gooseneck.[25]

The fall of the Roman Empire was followed by negligence and inability to fight fire. With the arrival of the Anglo-Saxons in England from the mid-fifth century, the situation changed dramatically. The wealth arising from inland resources made fire-fighting less important, as any destruction was followed by immediate reconstruction. Supply of materials was abundant and the need for fire-fighters lessened. As a result, Roman knowledge of fire-fighting was forgotten.[26] Fire-fighting methods and laws in England started to reappear only centuries later. In 872 a new law was written in Oxford, which demanded all house fires be put out with the ringing of the evening bell.[27] Special bell rings also announced certain events, such as the need for volunteers to fight fires. This method for calling for help was used in rural areas until the twentieth century. The fire bell of Sherborne Abbey in Dorset still has an inscription that says: 'Lord, quench this furious flame, Arise, run, help, put out the same.'[28] Following the Norman invasion in 1066, William the Conqueror imposed a stricter regulation in regard to putting out house fires at night. This was achieved by placing a metal cover over the open hearth, which is called *couvre feu* in French or *curfew* in English, and the ring of the bell was the curfew bell.[29] The modern understanding of the word curfew has a similar resonance, for as soon as it is imposed people have to stay indoors.

Organised fire-fighting re-emerged following the advent of industrialisation. By the early nineteenth century, most cities around the globe were partially destroyed by fire. The threat of urban fires, enhanced by bombing during periods of war, led towards the gradual creation and establishment of fire-fighting forces, initially on a private and local scale and, much later, on a national level. In Britain, immediately after the Great Fire of London in 1666, the fire-fighting service was provided by insurance companies and protected only buildings they were financially liable for.[30] Up until the end of the nineteenth century, it was the responsibility of each parish or town to provide fire-fighting forces and equipment. In 1707 in London, every parish was required to provide a fire engine and a leather pipe. Only in 1774, more than a century after the disaster, it was stipulated that 'every parish should provide three or more proper ladders of one, two or three storeys high, for assisting persons in houses on fire to escape there from'.[31] In 1832 the first union, called London Fire Engine Establishment, brought all the fire-fighters from insurance companies together with the primary concern of maintaining the protection of insured properties. A parallel body called Voluntary Society for the Protection of Life from Fire was established in 1836.[32] The peculiar dual system of fire-fighters and lifesavers was abolished at the command of Captain Sir Eyre Massey Shaw in 1865 with the enforcement of a relevant act, which established the Metropolitan Fire Brigade in London.[33] The London Fire Brigade was established a few decades later, in 1889, after a Local Government Act.[34]

Fire-fighting in the United Kingdom remained on a local scale until the beginning of World War II. Then, noting the catastrophic impact of the air raid bombings that took place during the Spanish Civil War and expecting a larger war to spread across Europe, the British Home Office urged for its reorganisation. In 1938, an Act of Parliament promoted the formation of a voluntary fire service, the Auxiliary Fire Service, the mission of which was to assist with existing fire-fighting forces.[35] Through the same initiative women joined the fire service with the creation of the Women's Voluntary Service for Air Raid Precautions.[36] In 1941, the National Fire Service was formed as a unified fire-fighting body. Later, in 1948, the fire brigade's administration returned to counties and local boroughs as this arrangement was considered more efficient and successful.[37]

Fire-fighting equipment

In response to the imminent threat of city fires, an additional process developed in parallel with fire-fighting. The fall of the Roman Empire was followed by the abandonment of fire-fighting equipment, which reappeared in 1548 after the German translation of Heron's work.[38] This translation instigated interest in forgotten mechanical fire-fighting methods, which in turn inspired the creation of the first fire engines. The first fire engine on wheels was invented by Anton Plater in Augsburg in 1548. After this a series of various designs was made, which required a constant supply of water from sources found in the vicinity. This was achieved with canvas or sailcloth buckets that were used to transport water to the engines. Yet the constant function of the fire engine demanded a constant supply of water. With the joint effort of people standing in a line, water was passed from the source to the engine, a task most frequently assigned to women, as men fought the fire on the frontline.[39] As well as their need for a constant supply of water, these early engines needed to remain a certain distance from the flames. They had to be close enough to be effective and far enough away that they did not burn. This issue was resolved in 1672 with the introduction of a leather hosepipe between the engine and a nozzle by two Dutch engineers, Nicolaes and Jan van der Heyden, which helped to keep the engine a safe distance from the fire.[40] The Heydens also helped solve the problem of supplying water with buckets. Initially they invented a canvas funnel on a wooden stand that could feed the engine with water through a pipe. Later, they invented a suction pump (fig. 6.5) that was connected directly to the engine.[41]

These early fire engines were also difficult to manoeuvre, as they were usually mounted on sleds and not wheels, so transport through narrow streets and alleys was problematic. Gradually fire engines increased in size, became more efficient and manoeuvrable, and started to be produced out of metal rather than wood. The first steam-powered pump was invented in 1829 by John Braithwaite and John Ericsson,[42] weighed 2¼ tons and pumped 30–40 tons of water per hour. It soon became highly unpopular as it replaced human labour. There was also general dissatisfaction because of their weight, the slow response time needed to physically transport them to fire incidents, and the fact that the fire of their boilers needed to be kept burning constantly so that the engine could pump water at any time.[43] Steam fire engines were later replaced by self-propelled ones, which were invented in 1862 in London by William Roberts.[44] They were three-wheeled, which allowed easy manoeuvres in narrow places, and their efficiency in throwing water was radically improved.[45] Later versions of fire engines were battery powered, easy to start, moved fast and provided adequate space on top of the vehicle for a person to drive. The last phase of fire engine design is the motor vehicle. Since 1910, motor vehicles have remained the basic layout of fire engines and today they differ only in the details.[46]

An equally interesting development that took place in parallel with the fire engine is the ladder. For hundreds of years, ladders were

Figure 6.5 Hose pump operated by members of the Sunderland Fire Brigade in the 1930s. © Mary Evans Picture Library.

used as a means of access and rescue in all buildings.[47] This remained unchanged until the nineteenth century when rescue from higher levels was needed.[48] The initial response to this need was the creation of the extension ladder, which was first designed in 1830 by a London-based artist called Abraham Wivell.[49] The need for further mobility led to the addition of carriage wheels. As the extension ladders became longer, this was followed by a demand for a rigid base. The German fire-fighting industry introduced ladders mounted on metal-frame beds that were supported on chassis. The ability to become longer demanded structural strengthening and this was achieved by trussing their length. Fire-fighting ladders were later produced with a hydraulically controlled, jointed and folding length with an operator's cage found at the top end (fig. 6.6). This improvement was welcomed and has been used since for aerial rescues.[50]

Figure 6.6 London Fire Brigade wheeled escape ladder and crew. The caption reads: London Fire Brigade motor escape competition G. District winners, Stoke Newington, 1925. © Mary Evans Picture Library.

The gradual increase in the use of plastic in households and certain types of industry was accompanied by fires that released highly toxic smoke. This rendered fire rescue exceptionally difficult and harmful. Efficient breathing apparatus became vital.[51] The first breathing aid of the late nineteenth century was a helmet made of rubberised material that sat on firemen's shoulders. The air was drawn in through a filtered mouth-tube and discharged through a valve. The first self-contained breathing apparatus was invented in the early twentieth century. It functioned with a closed-circuit oxygen supply that lasted for 30 minutes and later for one hour. One of the problems of combating fire using a breathing apparatus was the danger of disorientation in smoky environments, which resulted in firemen running out of oxygen supply. For this reason modern sets have a built-in safety margin, which can in principle give firemen enough time to escape a blazing building. However, if there is not enough time or if firemen find themselves in danger, breathing apparatus sets are equipped with audible alarms. The communication between firemen and with their base controls is also improved, as current breathing apparatuses carry reliable in-built radio sets.

Fire-fighting has been advanced by technological developments in other fields. Thermal imaging cameras (TIC), originally developed as night-time infrared army technology, are now widely used in fire rescues. These cameras can either be hand-held or helmet-mounted and assist firemen to find the exact location of a fire in smoke-filled environments. They are also used in tracing human casualties, as these cameras have the ability to detect the outline of bodies. Firemen's uniforms are made of fire-resistant and flash-resistant materials, which are commonly used in the space industry.[52]

An account of the evolution of fire-fighting technology reveals that it has acquired cyborg characteristics.[53] In *Simians, Cyborgs, and Women* the feminist theorist Donna Haraway compares late twentieth-century machines to earlier ones, explaining that the latter 'could not achieve man's dream, [they could] only mock it'.[54] Contemporary machines, on the other hand, 'have made thoroughly ambiguous the difference between natural and artificial, mind and body, self-developing and externally designed, and many other distinctions that used to apply to organisms and machines. Our machines are disturbingly lively, and we ourselves are frighteningly inert.'[55] Similarly, fire-fighting equipment is cyborgised in the sense that it comes to mediate between man and fire by reinforcing the former's physical abilities. In this light, rescue ladders and helicopters, breathing apparatuses, uniforms and cameras have evolved in such a way that they have become artificial extensions of man's hands, legs, eyes, lungs and skin respectively.

Fire-fighting and the city

The design of cities accommodates the demands of the fire brigade so that they can respond to fire incidents within a limited time and provide a functional and efficient fire-fighting operation. Urban design has evolved slowly and methodically throughout the years with the valuable contribution of specialists, who study the preparedness and responsiveness of the urban landscape in relation to fire risk.

In a letter dated 7 October 1905, the retired London Fire Brigade Chief Officer Captain 'R. N.' specifies the requirements for the location of fire hydrants (fig. 6.7) in relation to the risk associated with various properties:

> [I]n residential districts in which there are no large manufacturing premises, hydrants are fixed at a distance apart of about 450 feet

where the houses are less than 40 feet in height and situated adjacent to the footways. Where houses are more than 40 feet in height, hydrants are fixed at a distance of about 350 feet apart, and when houses are situated some distance from the footways the hydrants are fixed somewhat closer. In localities where large works, warehouses, shops, important business premises and public buildings are situated, the distance apart of hydrants varies in accordance with the nature of the property, and vary [on] average from 150 to 250 feet. Where there are exceptional risks, hydrants are fixed at closer intervals.[56]

The number of fire stations and their location in a city are not accidental. They are determined by a number of variables that respond to figures and data produced by fire brigades and governments (figs. 6.9, 6.10 and 6.11). In London, the Fifth London Safety Plan (LSP5) that was agreed in 2013 led to the closure of ten fire stations, as one of its

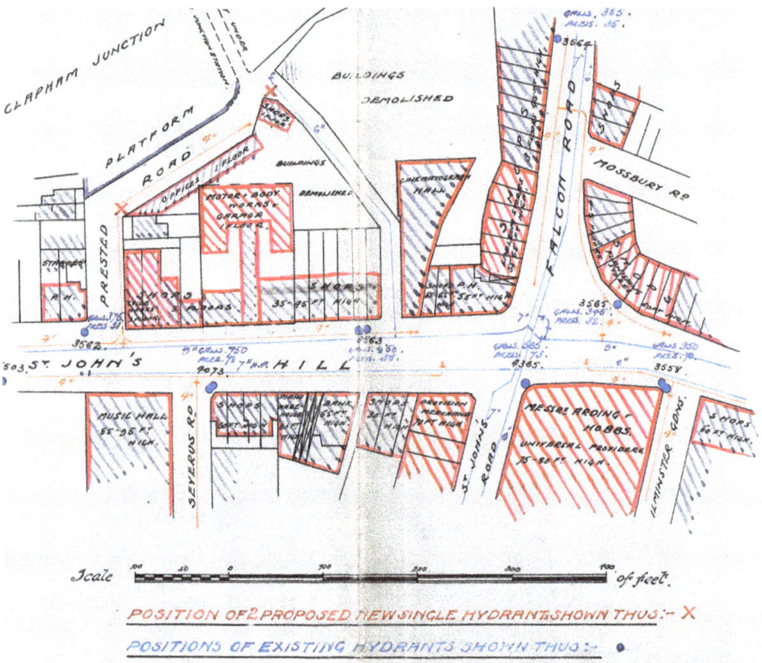

Figure 6.7 Map indicating the location of existing and proposed fire hydrants near Clapham Junction Station, London, 1912. © London Metropolitan Archives, City of London.

objectives was to save £29 million. Yet it is not only financial aspects that determine the fate of fire stations but also the overall strategy and performance of the fire service in response to fire. As soon as a fire is discovered, the fire brigade is usually contacted over the phone. This phone call is received by the fire brigade's control and despatch centres, which house the unseen protagonists of every emergency operation (fig. 6.8). Contemporary control centres are equipped with modern computer and communication technology, which allow the direction of every fire rescue operation to be conducted away from the noise, emotional charge and danger of onsite fire-fighting conditions.[57] As soon as the call is received, the address of the caller is displayed on a computer screen.

The call centre operator can automatically locate the nearest available fire station to the location of the incident. When its location is established, the fire station is informed about the fire. This information arrives with printed details of the address, the severity of the accident and the type of building that is involved.[58] On certain occasions, some control rooms are able to provide information, such as the layout of the burning building, location of the nearest fire hydrants and potential risks of the area, such as chemicals or explosives.[59] Permanently manned stations

Figure 6.8 London Fire Brigade Headquarters control room, 1939. © Mary Evans Picture Library.

Figure 6.9 London Fire Brigade map showing the location of fire stations in London, 1935. © London Metropolitan Archives, City of London.

Figure 6.10 Map of fire stations and fire alarms in the County of London, 1912. © London Metropolitan Archives, City of London.

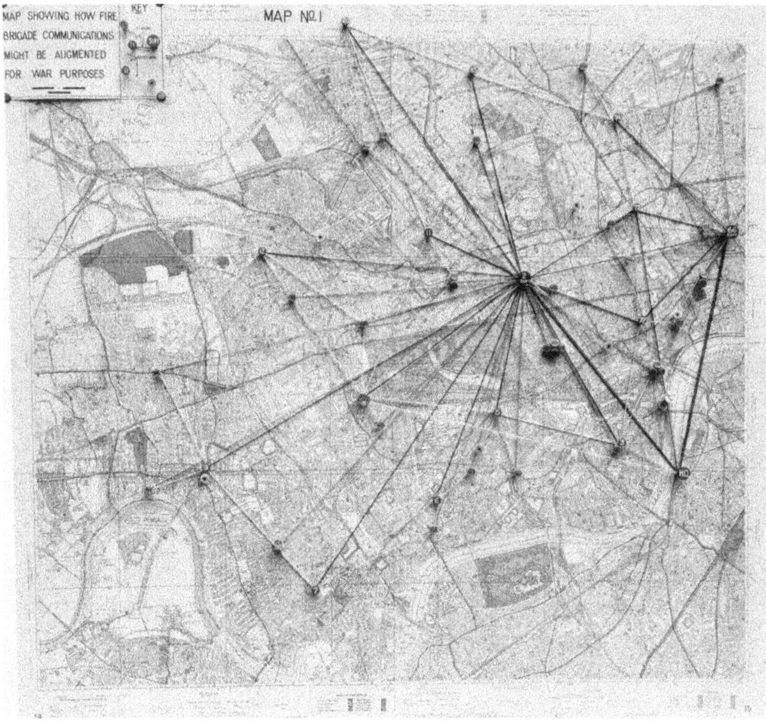

Figure 6.11 London Fire Brigade map showing how fire brigade communications might be augmented for war purposes, 1937. © London Metropolitan Archives, City of London.

are able to provide assistance within minutes after a call is made. For instance, the average response time to serious fires in England in 2016–17 was 8 minutes and 44 seconds. This time, which has increased over the last 20 years, depends on a range of factors, some of which are unpredictable, such as traffic levels or how long the control staff needs to assess the risk and attendance required at the fire incident.[60]

Reference to the development of the fire-fighting service, its technology and equipment demonstrates how the memory of fire catastrophes, and the possibility of future ones, informs and initiates processes that direct society towards a gradual and constant reorganisation. Since Roman times, fear of catastrophic fires has triggered the formation of groups of people who were given the responsibility of looking after civilians and taking action against fire. Following the fall of the Roman Empire, knowledge of fire-fighting was forgotten but later, due to severe urban fires that took place during the first years of industrialisation, organised fire-fighting re-emerged, and methods of combating the flames were reinvented. In the context of the current research, the emergence and development of fire-fighting can be perceived as an evolution that is not isolated but takes place in parallel with others, such as the evolution of fire-fighting technology and equipment, which is equally informed by the threat of fire. These interrelated and interdependent evolutions are additionally determined by the way a city is organised and they also inform the shape of the city. The fire engine, for instance, was initially immobile and wooden, whereas now it is adjusted on wheels and is fire-resistant. Its shape is adapted to fit through narrow alleyways and its ability to throw water has also improved. Equally, the city itself is spatially reconfigured as a response to the threat of fire. Contemporary cities are split into fire zones within which key sites, strategically chosen, are assigned to accommodate fire stations from which any fire incident can be dealt with within minutes. Fire hydrants are also tactically installed throughout the city so that they can offer a powerful supply of water to the fire brigade.

Fire protection and the building

Fire-fighting does not just take place outside the building. It is also absorbed into processes that take place within buildings. Buildings are protected against fire with fire protection systems that are classed as either active or passive. Most buildings feature an active fire protection system: objects or methods that respond actively to certain changes in the conditions of closed environments. Most commonly, this protection includes fire sprinkler systems mounted on ceilings, which are automatically activated as a result of an increase in temperature. The heat radiated by flames causes a glass component of the sprinkler system to detach itself and fall on the floor, followed by the distribution of water onto that particular area. The first automatic sprinkler system was invented in 1852 by William Macbay.[61] In 1861, Lewis Roughton created a more practical device, and in 1864 Major A. Stuart Harrison invented the first sensitive automatic sprinkler system. After World War I, two major developments took place: the introduction of the multiple control system and the glass bulb sprinkler head.[62] Detection systems are also part of an active fire protection strategy. Heat, smoke or flames can be detected by this equipment, which in turn activates an alarm that calls inhabitants for immediate evacuation. This system can be programmed to automatically inform the fire brigade and to trigger fire doors to shut and mechanical smoke vents to operate.[63] Fire extinguishers are also part of an active fire protection system. Their use is manual and dependent upon the inhabitants.

Active protection systems refer to certain technological devices, the mission of which is to protect the integrity of the building and provide safety to the inhabitants. Whereas in the past, the role of sprinkler or fire alarm was performed by humans (using the hose or hand-bell), now we 'employ technology … [as it] promises to bring forces of nature and culture under control, to liberate us from misery and toil, and to enrich our lives'.[64] The contemporary American philosopher Albert Borgmann, working with Heidegger's understanding of technology, understands the liberation and enrichment that technology brings about only in relation to the availability of goods. Accordingly, available technological goods contribute to our lives without imposing any burdens. For Borgmann, the hose or the hand-bell is a *thing* because one needs to put it into action to make it work. Things can only be understood in the context of their environment. They need physical engagement through the application of skill. *Technological devices*, on the other hand, which in this discussion are the sprinkler system or the fire alarm, perform a clear function that does not demand attention, engagement or skill, and because of this they come to be concealed or reduced in size.[65]

Passive fire protection deals with the overall design approach of a building with regard to fire and the appropriateness of the materials used for its construction. Fire compartmentation, for instance, is achieved by the employment of fire-rated building components such as walls, floors, doors and so on. Building components have to meet certain fire resistance standards, which in the unfortunate event of a fire have the ability to restrict the spread of fire and smoke within certain compartments.[66] This gives enough time for the occupants to escape the building safely and the fire brigade to conduct their work. Fire resistance standards are set and controlled by building and fire codes and regulations.

Fire escapes are also part of a passive fire protection system and they can be either internal or external to the building. Today most fire escapes are found within the shell of a building and they discharge at street level through fire exit doors. In the past, as soon as fire regulations were enforced for the first time, existing buildings had suddenly to comply and to reach an acceptable degree of fire safety. On certain occasions these adjustments were impossible, so fire escapes were mounted externally. A famous example is the steel fire escapes that can be seen extensively in the United States (fig. 6.12).

Figure 6.12 Steel fire escapes in Old City, Philadelphia. © Astero Klampatsa.

Fire-fighting methods and equipment, which were initially used to combat fire externally, were later adopted and adjusted so that they entered the building zone permanently, forming an integral part of a building's organisation. The fireman's hose was converted into an automatic sprinkler system or an internal hose reel, and the ladder into an external fire escape. One can argue that active fire protection systems lend cyborg qualities to architecture, as these systems cannot be overtaken by the natural fear of fire. In a recent paper on the concept of the cyborg, the hybrid between a machine and an organism, Matthew Gandy writes that cyborgs can be used to study 'the corporeal experience of space',[67] which he applies to contemporary cities to conceptualise their growth. Regarding the development of the contemporary home, he reflects on Anthony Vidler's argument[68] claiming that it 'has become a complex exoskeleton for the human body with its provision of water, warmth, light and other essential needs. The home can be conceived as "prosthesis and prophylactic" in which modernist distinctions between nature and culture, and between the organic and the inorganic, become blurred.'[69] Fire, heat and smoke detectors *sense* fire; fire alarms *react* and *communicate* with inhabitants to avoid loss of life and destruction of buildings. Automatic air vents allow the building to breathe and let the smoke out. Sprinklers complement firemen's duties while automatic fire doors compartmentalise the building and secure safe egress, thus enhancing the fire brigade's efforts in advance of their arrival. The fear of fire has equipped architecture with cyborg qualities that animate the building in the event of a fire.

The Grenfell Tower disaster

A few minutes before 1am on 14 June 2017, a fire broke out at Grenfell Tower, a 23-storey tower block in North Kensington, West London (fig. 6.13). The fire is believed to have started on the fourth floor, 'in and around'[70] a fridge freezer in the kitchen of Flat 16. The occupant of the flat, Behailu Kebede, was asleep in the living room and woken by the sound of the fire alarm as soon as it was activated. Kebede immediately contacted the fire brigade, which logged the call at 00.54. Five minutes later, the first fire engines were at the scene. The emergency crew reached Flat 16 13 minutes after the occupant contacted the emergency services. They first checked

Figure 6.13 Grenfell Tower in flames, 14 June 2017. © Natalie Oxford/CC BY 4.0.

both bedrooms and the living room before they entered the kitchen at 01.14. Daniel Brown, one of the firemen to enter the flat, describes how as soon as he entered the kitchen he encountered an 'isolated curtain of flame from about 2–3 feet in the air to the ceiling'.[71] Whilst the fire crew dealt with this fire, their thermal images reveal that embers were falling outside the kitchen window. It was clear by that time that the problem was much more serious than the fire in the single flat. According to evidence presented by the Irish forensic scientist Niamh Nic Daéid to the hearing at the Grenfell Tower Public Inquiry,[72] the fire had spread to the exterior of the building by 01.09, five minutes before the fire crew entered the kitchen.[73]

Once fire escaped outside the kitchen window, it spread rapidly through the apparently flammable cladding[74] and travelled upwards, first across the eastern face of the building and then across the northern one. A report produced by the fire engineer Barbara Lane discloses that 'the vertical fire spread primarily along column lines. The lateral fire was primarily along the cladding above and below window lines, the infill panels between windows, aided by the materials surrounding the windows and the materials surrounding the cavity formed between the new and old infill panels.'[75] Mobile phone footage reveals that the fire had reached the top of the eastern face of the tower by 01.26, less than 30 minutes after the fire brigade arrived at the scene.[76] At 02.06 London Fire Brigade declared the fire a major incident.

The tower was serviced by one stairwell and two fire-fighting lifts. The stairwell linked the roof with the ground floor and was equipped with 30-minute fire-rated self-closing doors. A 'stay put' policy was initially effective during the event. This non-evacuation policy of neighbouring flats was applied because the emergency services judged that the fire would not expand beyond the compartmentation zone of Flat 16 before it was fully extinguished. At 02.47, however, the stay put advice was abandoned. People remaining in the building at this point were trapped by flames and thick toxic smoke. The total number of victims confirmed by the authorities is 72 and police now assert that they will not find any more bodies. The opening statement of the London Fire Brigade during the Grenfell Tower Public Inquiry stated that this disaster was 'by far the most challenging incident which the London Fire Brigade (LFB) has experienced in living memory'.[77] It is the most severe tower block fire of the twenty-first century to date and has been compared to the Joelma Building fire disaster in São Paulo, Brazil, in 1974 that claimed the lives of 179 people.[78]

At the time of writing, the Grenfell Tower Public Inquiry is still underway but a leaked interim report prepared by the fire investigation experts BRE Global reveals specific 'calamitous deficiencies in the installation of the windows, cavity barriers and cladding system, and their failure to meet building regulations'.[79] Regarding the cavity barriers, the report discloses that their specified size was not sufficient to prevent the fire expanding to different sections of the building. The window frames were significantly smaller than the structural opening of the building whereas the space between frame and structure was filled by products, such as rigid foam insulation, that were not 30 minutes fire resistant. This allowed the fire to escape from the compartmentation zone of Flat 16 rapidly. The insulation used was flammable and carried no markings of the manufacturer. The cladding had a polyethylene (plastic) core that appears to be highly flammable. The front doors to many flats were not equipped with door closers, contrary to the demands of building regulations. The report also reveals that there was not a wet riser and a sprinkler system installed in the building, and that the single stairwell was 8 centimetres too narrow. Lastly, the landscaping around the tower did not allow more than one fire engine to attend at the base of the tower.[80]

The Grenfell Tower disaster confirms how easily a small issue with an electrical appliance can cause a real catastrophe. A little spark can develop into an uncontrolled fire and spread rapidly throughout a building. This major accident can be scrutinised through various angles that clearly challenge the efficiency, enforcement and compliance of fire regulations, the fire resistance of construction materials and the adequacy of fire resistance testing, and the preparedness (and perhaps also the ability) of the fire brigade to deal with accidents of such a scale. Yet as the leading housing law barrister in the United Kingdom, Andrew Arden, points out, the Grenfell Tower disaster also highlights the overall British approach to social housing. He explains that during the 1960s social housing was underfunded, and although the structural integrity of tower blocks was secured, there were neither thermal insulation requirements nor fire protection legislation tailored to this building typology. The lack of thermal insulation requirements, Arden continues, is not irrelevant to the fire accident, as the cladding that was later added aimed at improving the existing thermal performance of the tower.[81]

Fire and building regulations

Post-Grenfell Tower reflections confirm that the enforcement and compliance with building regulations, from the conception of a design to the construction and maintenance of the building, are crucial not only for the integrity of the building but also for our safety. Building regulations have existed for a long time. The first recorded evidence is the rules imposed by the Babylonian king Hammurabi in 1750 BC. Articles 229 to 235 deal with design issues. Article 229 specifically states that 'if a builder builds a house for someone, and does not construct it properly, and the house which he built falls in and kills its owner, then that builder shall be put to death'.[82] At first, building regulations dealt with the protection of people occupying buildings; only in more recent years have these rules included the protection of the property itself.

A significant step towards the formation of legislation in the United Kingdom was the City of London's code established in 1189 called the Assize of Buildings, which enforced the construction of common walls, rights to light access, drainage and safe egress in case of fire.[83] In Siena, Italy, building laws were introduced in 1262, and by 1309 any new house required planning permission prior to its erection.[84] Historically, however, the main reason that drove societies towards the establishment of building rules was the catastrophes caused by fire. After the Great Fire of London in 1666, the building legislation of the city became very strict and complex. Amongst these restrictions, stone and brick[85] became compulsory materials, and tiled roofs replaced thatched ones.

The most sudden and rapid demand for building regulations was observed during the second half of the nineteenth century as a result of the Industrial Revolution. The success of James Watt's invention of the steam engine brought about the demand for new architectural solutions. Until then, except at seaports, industrial goods were transported in small quantities, and were stored and merchandised in small shops. The invention of the engine allowed mass transportation of goods to city centres. New building solutions were needed, such as large open-plan and multi-storey warehouses. At that time the main fuel for city fires was the building itself, but following the Industrial Revolution, the contents of industrial buildings became the predominant threat.[86] In addition, the migration wave from rural areas to cities led to the creation of high-density urban centres. This developing new urban reality was accompanied by the constant threat of fire and epidemics. The increasing number of new buildings demanded a shift of liability from property owners to their lenders and insurers. The protection of city dwellers and their property was thus an urgent issue, which pushed towards the creation of building legislation. In the United States after the fire in Chicago in 1875, the first American building code was established. Following Chicago, other cities in the United States wrote their own local building rules but they lacked the knowledge and experience to be obtained from the aftermath of fire, and fell short of specialised resources needed to apply to the task. The catastrophic fire in Baltimore in 1904 urged the creation of a model code. In 1905 the National Board of Fire Underwriters, an insurance industry group, compiled the first National Building Code in the United States.[87]

Building regulations are an area of constant research. They are composed by experts whose task is to gather information about the causes and impacts of failures and accidents that take place in the built environment. Today building regulations are an advanced and complex reality for which the majority of professionals from the building industry feel great disrespect. Designers in particular perceive them as a body of unpleasant restrictions that condemn freedom of expression and innovation.[88] The impact of fire seems to interest a limited number of specialists, usually insurance experts and surveyors acting on behalf of insurance companies.

The current legislation presiding in England is the Building Regulations 2000, which were designed by the Secretary of State for the Environment under powers delegated by Parliament under the Building Act 1984.[89] The rules on fire safety, to be found in 'Approved Document Part B', form the longest and strictest part of the Building Regulations. This document provides fire safety guidance for the construction and maintenance of all types of building, from low-rise dwellings to high-rise office spaces. Fire safety compliance is essential but there are other routes, apart from compliance with the Approved Document Part B, which one can take. A building can also be designed following the stipulations of British Standards or the computational assessments of a fire-engineering approach. Although all three methods are equally acceptable, their prescriptions might vary, which justifiably raises questions about safety.

Fire regulations are principally tailored for a single purpose: the containment of fire in a building and the protection of its occupants. Nevertheless there are some parameters that cannot be predicted or avoided. Fire regulations cannot control the flammability of contents brought into a building by the occupant. In this case fire regulations are destined to protect the neighbours.[90] Commenting on a tragic fire accident at Lakanal House in Camberwell, London, on 3 July 2009 that caused six people to lose their lives, the architect Sam Webb adds that the lack of cleanliness is an additional threat that needs to be taken into consideration when thinking about fire safety.[91] The lack of cleanliness can cause an unpredictable spread of fire and can take everyone by surprise. This issue was initially addressed after the fierce fire at King's Cross Station on 18 November 1987, which probably started because of a burning match or cigarette and killed 31 people. According to the London

Fire Brigade, during the annual inspection the station was found to be in a substantially dirty state, providing storage for combustible materials.[92] Similarly, Webb explains, the fire at Lakanal House would not have spread so rapidly if the public areas of the building were kept clean – hence his insistence on the enforcement of relevant legislation.[93]

Another unpredictable and challenging parameter is the panicked reaction of people in the presence of fire. The concept of panic has a long history[94] and refers to a contagious feeling that is not only limited to fire. When faced with fire, the reaction of people is in most cases unpredictable, as panic overcomes the human capacity for rational thinking.[95] According to records of fire incidents in the nineteenth century, the buildings that created the biggest scenes of panic were theatres (fig. 6.14). These were the most common venues for public entertainment yet their design incorporated no safety measures.[96] Between 1800 and 1875 there were 46 recorded theatre fires in the United Kingdom, which were caused by the gas lamps used to light up the stage. Due to the high number of these accidents, emotions of panic were common when people gathered in big and crowded public buildings. The fact of panic attending a fire was gradually absorbed and encoded in regulations. Today it manifests itself through compulsory design implementations, such as easily identifiable

Figure 6.14 Aftermath of a fire at the Arts Theatre Club, Great Newport Street, Central London, 4 August 1951. © Mary Evans Picture Library.

fire-resistant escape routes of certain widths, and controlled provision of emergency lighting and ventilation.[97]

The enforcement of and compliance with building regulations can be perceived as an act of compromise to an ideal that looks towards the creation of a fireproof world. The possibility of the total elimination of destructive fires, whilst fire is still in the service of humans, is probably an unrealistic scenario. This would demand the exclusive use of non-combustible materials, which would in turn imply an urban environment built anew. The aspiration towards a fireproof world made of non-flammable materials can be seen in the technology of building materials, as, for example, in the extensive use of asbestos.[98] In the late nineteenth century, asbestos became increasingly popular within the building industry due to its fire-resisting properties.[99] Evidence for its use can be traced back to the seventeenth century in what is called the salamander's wool, a type of woven cloth that was thrown into fire and was later emerged intact. The fascination with the use of asbestos, apart from its capacity to resist heat, electricity, acidity and dampness, is also attributed to the fact that it can be woven like cotton; it can even be turned to fibre and powder with remarkable tensile strength. The first time that asbestos worked efficiently as a fire-resistant material was during a fire incident at the Queen's Theatre in Manchester in 1890, at a time when asbestos was widely used in safety curtains. During that fire, asbestos curtains managed effectively to divide the theatre[100] and prevent the expansion of flames to subsequent spaces.[101] This can be understood as an early version of fire compartmentation, which today is achieved with the use of fire doors and curtains.

Through a critical analysis, building regulations appear to function like an archive, occupying a space in which a codified memory of accidents and destruction is securely stored. The contents of this archive – the building regulations – inform architecture through their enforcement, which is what Derrida calls the archive's violence. The traumatic memory of accidents and destruction is *violently* externalised and embodied in the end product, in architecture, and manifests itself through the design, method of construction, materiality and so on. This archive is usually found in a static state until an accident occurs. This is when the archive is reshuffled and updated with new information. Architecture functions like an archive, as it carries a frozen memory of previous catastrophes through design that abides by building regulations. In the case of an accident, the architectural archive is put into motion and informs regulations through what is left behind, its ruins or ashes. In this sense, architecture's own catastrophe informs architecture itself in a vicious circle.

Building regulations and creativity

The restrictions imposed by the enforcement of building regulations can lay the ground for something new to emerge. Through legislative restrictions the building process enters previously uncharted territory. According to the ancient Greek concept of *horismos* (χωρισμός, dividing), as soon as one sets boundaries or restrictions, something does not stop but actually begins to emerge. The Party Wall Project, an experimental project by Mae Architects in collaboration with Slider Studio, illustrates this idea vividly (fig. 6.15).

This project is inspired by the compulsory use of party walls in the urban fabric. The construction of party walls was enforced for the first time after the Great Fire of London in 1666. Since then, a thick common wall (party wall) has separated adjoining properties with the intention to prevent the spread of fire and to protect the immediate neighbours.[102] The enforcement of party walls has created a unique urban grid and buildings can only emerge within the boundaries defined by it. Considering the spatial restrictions imposed by this urban grid and focusing on the typology of terraced housing, Mae Architects employed Slider Studio's computational simulations to explore a countless number of possible design

Figure 6.15 Aspect of the Party Wall Project by Mae Architects and Slider Studio, 2005. © Mae Architects and Slider Studio.

configurations through mixing materials, forms, functions, sunlight conditions, colours and so on. This project demonstrates that restrictions are not limiting but instead create an environment that carries the potential of fresh and radical creativity.

The restrictions of building regulations can equally be perceived as a positive and liberating tool in the hands of the architect. A project that celebrates this argument is the Brick House by Caruso St John Architects, a shortlisted nominee for the Stirling Prize in 2006.

The Brick House in west London (figs. 6.16 and 6.17) stands out as a remarkable design that is generated as a compromise to various restrictions. The site is surrounded by three taller buildings, and has an awkward shape and unconventional access. Due to planning restrictions, this residential building could only raise one storey up and could not have direct views to three sides. In response to these restrictions, Caruso

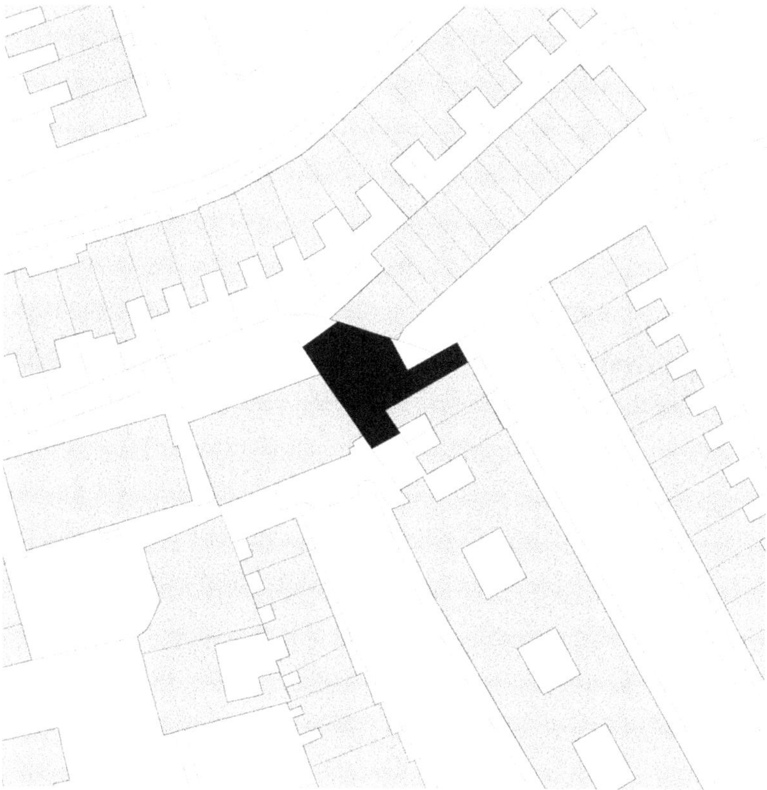

Figure 6.16 Site plan of the Brick House by Caruso St John, 2006. © Caruso St John Architects.

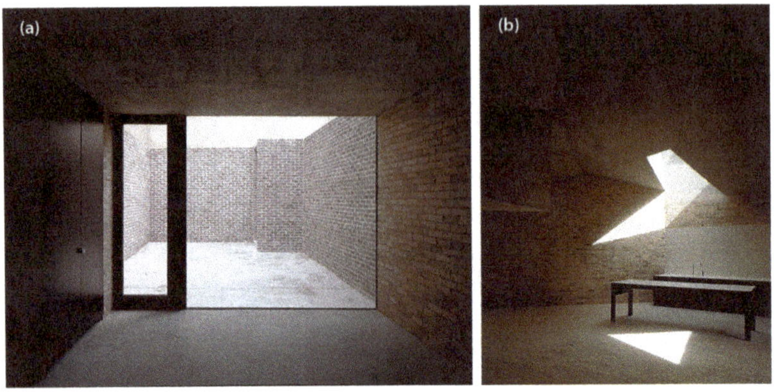

Figure 6.17a, b Interior images of the Brick House by Caruso St John Architects, 2006. © Hélène Binet.

St John Architects designed a cave-like space where sunlight is captured through shafts and projected onto the ceilings. The rooms are placed on the lower ground level and have direct views to a small internal courtyard. This compromised design, although foreign to conventional typologies, is a remarkable illustration of how the enforcement of legislative restrictions has the potential to inspire great creativity.

Research on fire prevention

The imminent risk imposed by the flammability of the contents of buildings has led to the development of specialised research on fire prevention, which calculates the behaviour of combustible materials in relation to fire. Fire prevention deals primarily with the calculation of the total heat of combustion, which is the total thermal energy per unit mass that is emitted through complete combustion. It also deals with the rate of heat release, that is, the intensity of the material's output of thermal energy per unit mass per unit time. The calculation of exothermic properties allows for the estimation of the material's contribution to a fire scenario. Research on fire prevention also looks at the endothermic properties of materials, the calculation of which refers to what is needed for sustaining combustion (heat capacity, sensible heat, latent heat).

Thermal conductivity, which is the rate at which heat enters from the surface into the interior of the material to allow the process of decomposition to take place, is another crucial aspect in this type of research.[103] Fire researchers reconstruct fire incidents in laboratories to monitor the performance of combustible materials. This test, however, represents only one combination of conditions that might take place in a real fire, as its development in space is always unpredictable. Therefore fire test design requires the consideration of many variables. The most common technique developed by researchers is the *tunnel test*. The material to be tested is placed on top of a long rectangular passageway, on one end of which a large fire is introduced for a certain period of time. The material in question supplies fuel to the burner and effectively allows the fire to progress further down the tunnel. The propagation of fire inside the tunnel and the distance that it travels until it burns out is recorded and its measurement represents the degree of flammability. If the result of the tunnel test is below a certain degree, the material is rated as non-combustible.[104] In 1950, the tunnel test was established as the standard method to predict the way fire spreads, particularly on flooring materials. This method proved to be misleading, however, as the performance of the same materials in real fires was often different from the one predicted.[105]

Fire tests can only reliably establish the physical properties of a material. Real fires exhibit quantitative and qualitative differences that cannot be taken into consideration in a test, such as heat flux, airflow and so on. Hence the establishment of research centres, which have the capacity to house full-scale fire experiments. These experiments are expensive but their significance is increasingly being recognised.[106] The first fire laboratory in England was established after a decision of the British Standards Institution in 1932 in Elstree. It was the third laboratory to operate internationally after two centres in Chicago and Stockholm. The results emerging from fire laboratories are particularly informative. Structures are carefully subjected to measured pressures and then heated to very high temperatures, from 1000 to 2200°C. The experiment finishes with the structure's resistance to water, as would occur in a real fire incident with the water introduced by the fire brigade. A nine-inch brick, for example, is expected to resist a six-hour combustion under a 40-ton pressure and a subsequent six-minute cooling down with cold water. The purpose of these laboratories is not only the observation of materials under combustion but also the testing of fire-fighting techniques and extinguishing appliances.[107]

The expensiveness of full-scale fire experiments has led researchers to the creation of an alternative method of testing, which is cost-effective but scaled down.[108] The National Bureau of Standards in the United States has validated 1:4 as the acceptable scale for testing. Although the results from this test are precise, there are certain parameters that cannot be scaled down. For example, the height of flames cannot be scaled down proportionally. This means that the fire damage measured in small-scale experiments is less severe than in full-scale ones yet the final results can be adjusted accordingly.

Alongside real fire experiments, researchers have also turned their interest towards the use of mathematical models. Early attempts at mathematical fire models were modest. They tested basic spaces with a single-fuelled fire burning in the middle of the room, whereas most parameters of a real fire event were not considered. These models gradually started to become more complex by subdividing the volume of air in the room into smaller units. The advance of computer technology in the mid-1970s allowed models to be accomplished in more detail, including parameters such as different room sizes and geometries.[109] Today computational fire modelling is a method widely used by fire engineers. It has advanced to such an extent that it can predict the turbulent flow of gases inside a building as well as keep track of the changing positions and sizes of soot particles.[110] Computational fire modelling is able to predict the duration of a fire developing in any given space from ignition to the point of its complete exhaustion. The calculation takes into consideration all the potential sources that can fuel a fire, and according to crosschecks with results obtained through full-scale experiments the accuracy of these models is plus or minus 10 per cent. This margin of error is critical when it comes to calculating the time needed for escape during an emergency. It is thus debatable how much confidence one can have in computer simulations. There are parameters that the software can easily and quickly calculate, yet others, such as the panic-stricken behaviour of people escaping from fire, remain mere predictions.[111]

Research on fire prevention entails exploring and predicting the way fire develops in space and how materials burn. Considering that the aftermath of real fire accidents does not contribute an adequate amount of information to the study and avoidance of similar future events, fire researchers conduct real life fire experiments in laboratories or simulate fire incidents in buildings. Architecture is reproduced either in smaller physical scale or digitally, and is then set on fire. For the French philosopher Bruno Latour, experiments that take place in laboratories are understood as the following:

> The guinea pig alone would not have been able to tell us anything about the similarity of endorphin to morphine; it was not mobilisable into a text and would not help convince us. Only a part of the gut, tied up in a glass chamber and hooked up to a physiograph, can be mobilized in the text and add to our conviction.[112]

The results of experiments are delivered in the form of scientific texts. Questioning the source of these texts, however, Latour asks: 'What is behind a scientific text? Inscriptions. How are these inscriptions obtained? By setting up instruments.'[113] The inscription of scientific results forms the written memory of experiments. Yet what follows after the scientific text is in hand? With regard to fire experiments, this text informs architecture, as it decides upon the building's materiality, structure, contents etc. In other words, the memory of destruction of the physical or digital replication of architecture is archived in the end product, in architecture itself.

Architecture on fire: A Bachelardian approach

As discussed in the preceding paragraphs, contemporary societies make a thorough and elaborate investment in protecting their members from fire as well as in preserving the integrity of buildings. This investment manifests itself in various ways. Buildings are designed and equipped to combat fire at all stages of combustion. Smoke detectors respond to traces of smoke, which are products of the smouldering stages of combustion or incandescence. Heat and fire detectors respond to later stages of combustion when flames have already developed. Part of their response to fire is the activation of the sprinkler system, which emits water into the building from ceiling outlets. At the same time an alarm warns inhabitants to evacuate immediately, while an automatic signal is sent to the fire brigade to alert them to the incident. The evacuation of a building takes place through designated fire routes and exits, which are designed in such a way as to provide a safe and quick egress free of smoke and flames for a period of time. This is achieved through fire compartmentation and the use of fire-resistant materials.

This approach to fire protection can be explored through Bachelard's views on duration and evolution. Bachelard, opposing Bergson's temporal continuity, perceives evolution as discontinuous: its duration is broken down into points and intervals. The duration of combustion in a building is similarly broken down into different stages. Starting from the smouldering stage, it gradually escalates until it reaches full combustion. Architecture and fire protection technology reflect this discontinuity by responding to each stage of combustion accordingly. Smoke detectors detect the early stages of combustion whereas heat detectors detect later ones. The positioning of fire exits in a building responds to a discontinued evolution. The duration of escape from any point in a building to a safe place outside cannot exceed a certain amount of time. This duration is spatialised as it is calculated in distance. Abiding by building regulations, the positioning of fire exits punctuates the duration of escape to the maximum distance possible from any given point inside the building.

The way fire is combated is the result of an ongoing, organised attempt to avoid recurring accidents. If one considers architectural evolution in Bachelardian terms, a fire catastrophe is the temporal death of architecture. This temporal death punctuates architectural evolution, it makes it discontinuous, and at the same time it indicates an instant of creation, as a new building or part of a city is usually erected to replace the destroyed one. The end product of combustion, the destroyed building, can be perceived as a body of useful knowledge that informs fire research and technology. For Bachelard, knowledge 'taken at the moment of its constitution is polemical knowledge; it must first destroy in order to make room for its constructions. Destruction is often total and construction never completed.'[114] The knowledge attained through the study of fire accidents agrees with his notion of knowledge production. Knowledge is the product of the dialectic between theory and practice where 'concepts need to become technical to be precise and to develop, and technical applications are performed according to theoretical procedures and on scientific objects, that is, objects created by science'.[115] In this sense a destroyed building is the theoretical coordinate upon which new technologies are tried out and through which new knowledge is produced.

Architecture's destruction provides useful knowledge that informs the revision of building regulations. This knowledge, which essentially forms a memory of destruction, informs future architectural creations. Through the enforcement of building regulations, the memory of destruction is externalised in the design, materials and building technology. Building regulations filter useful knowledge of the past, as 'what remains of the historical past … [is] only that which has reasons for beginning again'.[116] This externalisation secures memory, as 'the more securely … [memories] are fixed in space, the sounder they are'.[117] A Bachelardian understanding of architecture's response to destruction demonstrates that architecture absorbs and reflects the discontinuity of its own evolution. A point that punctuates architecture's evolution is its destruction, its temporal death. Architecture operates as an archive in which a reduced memory of its entire past is stored and recorded. The moment of its destruction signifies a point in time when existing knowledge is reorganised. New buildings only carry knowledge that has been previously proven to be useful. The rest detaches itself from architecture's evolution and becomes part of history. In this sense architecture operates as an archive that updates and reorganises itself through its very destruction.

Architecture on fire: A Bergsonian approach

The multifaceted responses to the imminent threat of fire, if considered in Bergsonian terms, make up a wider, more detailed picture of the same evolution, the evolution of architecture, and one where destruction only indicates the temporal aspect of this evolution. Bergson's evolution is a dynamic, open-ended process that unfolds constantly. This evolution is creative and unpredictable and it takes place in duration, in real time, which he defines as 'the continuous progress of the past which gnaws into the future and which swells as it advances'.[118] In order to understand the concept of Bergsonian evolution, one must first make a distinction between instinct and intelligence. The duration of evolution, Bergson explains, must be experienced instinctively, yet it is perceived with our intellect. The intellect has a tendency to concentrate and depend upon inert matter. And even if it does consider living matter, it still treats it as if it were inert. Any fluidity or movement escapes it, and what it perceives instead is a series of discontinuities, of immobilities, that give the impression of continuity and duration. Accordingly, the partial or total destruction of architecture is merely a snapshot of architecture's overall evolution.

Both instinct and intelligence are tendencies, which in earlier stages of evolution used to interpenetrate each other. Now they only hold traces of their common roots. Instinct and intelligence possess innate knowledge. The former carries innate knowledge of things, of matter, and the latter of relations: the knowledge of how things relate to each other, of their form.[119] In *Creative Evolution*, Bergson makes a further distinction between instinct and intelligence in terms of how inventive these two tendencies are. Invention is complete as soon as there is a material end product, the creation of an instrument.[120] How these instruments are manufactured and used discloses a fundamental difference between instinct and intelligence: 'Instinct perfected is the faculty of using and even of constructing organised instruments; intelligence perfected is the faculty of making and using unorganised instruments.'[121] In this sense, Bergson explains, an insect or an animal uses an already assembled instrument that it possesses and is part of its body, such as a nose, eye or wing, to perform its task. This is an instinctive function. Conversely, humans, operating with their intelligence, assemble unorganised instruments that are external to their bodies and manufactured for specific purposes. For this precise reason an instrument operated instinctively is perfect for its function because, if its use were modified, the species itself would be different. An intelligently manufactured instrument, however, is imperfect. As Bergson points out, the instrument 'is made of unorganised matter, it can take any form whatsoever, serve any purpose, free the living being from every new difficulty that arises and bestow on it an unlimited number of powers'.[122] An instrument created by intelligence cannot be restricted merely to its primary function. It can be used in different ways and for various reasons. It allows for unlimited improvement and application. Yet the only advantage of intelligence over instinct is to be found in later stages of evolution, when intelligence 'proceeds to construct constructive machinery. At the outset, the advantages and drawbacks of the artificial instrument and of the natural instrument balance so well that it is hard to foretell which of the two will secure to the living being the greater empire over nature.'[123]

In any case, the fact that intelligence constructs and makes use of unorganised instruments reveals an additional aspect of human evolution, which is of major importance. Evolution does not only embrace the organic changes that take place in our physical nature but also includes all developments in our intellectual and material culture. In the context of this book, one can draw parallels between human evolution and the evolution of material culture in regard to fire accidents, such as fire-fighting equipment, technology and so on. Fire accidents, which one would normally perceive as obstacles to architectural evolution, are dealt with by our intelligence. We employ external tools that we ourselves have invented to overcome these obstacles. When a fire breaks out, the sprinkler system replaces the manual labour of fire-fighters by throwing water on the flames. The smoke and fire detectors replace our senses of smell and touch respectively, and fire alarms announce the event. Accordingly, as architectural evolution unfolds, it initiates other evolutions that are not only parallel to but also dependent on it.

A Bergsonian understanding of evolution also implies that at every moment of its development, this dynamic, ever-changing process archives all knowledge related to intellectual and material culture. As this knowledge is a product of a process in flux, it never stays static. The evolution of knowledge also takes place in duration but, as it is our intellect's tendency to perceive fluidity as discontinuous, it breaks down its trajectory into immobilities and projects knowledge, as if it were static, onto the form of things. Therefore the knowledge we obtain by reflecting on any formal arrangement is just a fraction of the whole knowledge that is inherent to evolution, as 'what is real is the continual change of form: form is only a snapshot view of a transition'.[124]

Notes

1. Rome was totally destroyed by fire in 64 BC. Large parts of cities have also been destroyed by fire, as in the Great Chicago Fire in 1871, the Great Fire of London in 1666, the San Francisco earthquake fire in 1906 and the fire in Tokyo in 1923.
2. Paul Stollard and John Abrahams, *Fire From First Principles: A Design Guide to Building Fire Safety* (London and New York: E & F. N. Spon, 1991), 5–6.
3. Arthur E. Cote, *Fundamentals of Fire Protection* (Quincy, MA: National Fire Protection Association, 2004), 107.
4. Aqua Group, *Fire and Building: A Guide for the Design Team* (London: Granada, 1984), 2.
5. Aqua Group, *Fire and Building*, 2–4.
6. James G. Quintiere, *Fundamentals of Fire Phenomena* (Chichester: John Wiley & Sons, 2006), 3.
7. Aqua Group, *Fire and Building*, 3.
8. John W. Lyons, *Fire* (New York: Scientific American Books, 1985),104.
9. Aqua Group, *Fire and Building*, 182–3.
10. Lyons, *Fire*, 85.
11. William K. Bare, *Fundamentals of Fire Prevention* (New York: Wiley, 1977), 123.
12. Stollard and Abrahams, *Fire from First Principles*, 29.
13. Aqua Group, *Fire and Building*, 9.
14. Stollard and Abrahams, *Fire from First Principles*, 11.
15. Stollard and Abrahams, *Fire from First Principles*, 13–14.
16. Aqua Group, *Fire and Building*, 13–14.
17. Pyne, *Fire: A Brief History*, 109.
18. David Burgess-Wise, *Fire Engines and Fire-Fighting* (London: Octopus, 1977), 7.
19. This name was probably given to this force because, acting like parents of the public, they took responsibility for the latter's safety.
20. G. V. Blackstone and Fire Protection Association, *A History of the British Fire Service* (Moreton in Marsh: Fire Protection Association, 1996), 1.
21. Blackstone and Fire Protection Association, *A History of the British Fire Service*, 2.
22. E. Green-Hughes, *A History of Firefighting* (Ashbourne: Moorland, 1979), 11–12.
23. Arthur Ingram, *A History of Fire-Fighting and Equipment* (London: New English Library, 1978), 7.
24. Burgess-Wise, *Fire Engines and Fire-Fighting*, 6.
25. Ingram, *A History of Fire-Fighting and Equipment*, 9.
26. Burgess-Wise, *Fire Engines and Fire–Fighting*, 11.
27. Green-Hughes, *A History of Firefighting*, 15.
28. Blackstone and Fire Protection Association, *A History of the British Fire Service*, 7.
29. Pyne, *Fire*, 109.
30. W. E. Jackson, *London's Fire Brigades* (London: Longmans, Green and Co., 1966), 11.
31. Jackson, *London's Fire Brigades*, 28.
32. Green-Hughes, *A History of Fire-Fighting*, 34.
33. Green-Hughes, *A History of Fire-Fighting*, 57.
34. J. B. Nadal, *London's Fire Stations* (Huddersfield: Jeremy Mills, 2006), 99.
35. London Fire Brigade, 'Firefighters of the Second World War', https://www.london-fire.gov.uk/museum/history-and-stories/firefighters-of-the-second-world-war/, accessed 26 September 2018.
36. Royal Voluntary Service, 'Our History', https://www.royalvoluntaryservice.org.uk/about-us/our-history, accessed 26 September 2018.
37. London Fire Brigade, 'Firefighters of the Second World War'.
38. Ingram, *A History of Fire-Fighting and Equipment*, 8.
39. Ingram, *A History of Fire-Fighting and Equipment*, 9–10.
40. Burgess-Wise, *Fire Engines and Fire-Fighting*, 13.
41. Ingram, *A History of Fire-Fighting and Equipment*, 11.
42. Roger Pennington, *British Fire Engine Heritage* (London: Osprey, 1994), 21.

43. Ingram, *A History of Fire-Fighting and Equipment*, 14.
44. Bart H. Vanderveen, ed., *Olyslager Auto Library: Fire-Fighting Vehicles 1840–1950* (London and New York: Frederick Warne, 1976), 9.
45. Burgess-Wise, *Fire Engines and Fire-Fighting*, 75.
46. Ingram, *A History of Fire-Fighting and Equipment*, 31.
47. Fire Department, *Manual of Firemanship: A Survey of the Science of Fire-Fighting*, book 5, *Ladders and Appliances* (London: Her Majesty's Stationery Office, n.d.), iii.
48. K. R. M. A. Gilbert, *Descriptive Catalogue of the Collection Illustrating Fire Fighting Appliances* (London: Science Museum, 1969), 22–3.
49. Green-Hughes, *A History of Fire-Fighting*, 34–5.
50. Ingram, *A History of Fire-Fighting and Equipment*, 71–3.
51. Fire Department, *Manual of Firemanship: A Survey of the Science of Fire-Fighting*, book 6, *Breathing Apparatus and Resuscitation* (London: Her Majesty's Stationery Office, n.d.), iii.
52. Neil Wallington, *Fire-Fighting: A Pictorial History* (Bristol: Parragon, 1997), 40–1.
53. Etymologically, the word 'cyborg' derives from the word 'cybernetics', which was for the first time coined in the 1940s by the mathematician Norbert Wiener, and the word 'organism'. The term 'cyborg' was initially used by NASA scientists and referred to 'a series of experiments that explored how the human body might be technologically enhanced in order to allow space travel'. Matthew Gandy, 'Cyborg Urbanization: Complexity and Monstrosity in the Contemporary City', *International Journal of Urban and Regional Research*, 29: 1 (March 2005), 27, note 1.
54. Donna Jeanne Haraway, *Simians, Cyborgs, and Women: The Reinvention of Nature* (New York: Routledge, 1991), 152.
55. Haraway, *Simians, Cyborgs, and Women*, 152.
56. Letter, London Fire Brigade Chief Officer Captain R.N. (Rtd.), 7 October 1905, LCC/FB/GEN/06: Water supplies, London Metropolitan Archives.
57. Fire Department, *Manual of Firemanship: A Survey of the Science of Fire-Fighting*, book 10, *Fire Brigade Communications and Mobilising* (London: Her Majesty's Stationery Office, n.d.), 25.
58. Fire Department, *Manual of Firemanship: A Survey of the Science of Fire-Fighting*, book 10, 50–2.
59. Wallington, *Fire-Fighting: A Pictorial History*, 50–1.
60. Home Office, 'Response Times to Fires Attended by Fire and Rescue Services: England, April 2016 to March 2017', 25 January 2018, https://www.gov.uk/government/statistics/response-times-to-fires-attended-by-fire-and-rescue-services-england-april-2016-to-march-2017, accessed 20 June 2018.
61. Fire Department, *Manual of Firemanship: A Survey of the Science of Fire-Fighting*, book 9, *Fire Protection of Buildings* (London: Her Majesty's Stationery Office, n.d.), 3–4.
62. Fire Department, *Manual of Firemanship: A Survey of the Science of Fire-Fighting*, book 9, 4–5.
63. Aqua Group, *Fire and Building*, 66–7.
64. Albert Borgmann, *Technology and the Character of Contemporary Life: A Philosophical Inquiry* (Chicago: University of Chicago Press, 1984), 41.
65. Borgmann, *Technology and the Character of Contemporary Life*, 42.
66. HM Government, *The Building Regulations 2010*, Fire Safety, Approved Document B, Volume 1: Dwellinghouses (2006 edition incorporating the 2010 and 2013 amendments) (London: NBS, 2011), 57–65.
67. Gandy, 'Cyborg Urbanization', 28.
68. Anthony Vidler in his essay 'Homes for Cyborgs' explains that the boundaries between nature and technology are now blurred, as 'the body, itself invaded and reshaped by technology, invades and permeates the space outside, even as this space takes on dimensions that themselves confuse the inner and the outer, visually, mentally, and physically.' Anthony Vidler, *The Architectural Uncanny: Essays in the Modern Unhomely* (Cambridge, MA: MIT Press, 1992), 147.
69. Gandy, 'Cyborg Urbanization', 28.
70. The beginning of the fire 'in and around' the fridge freezer was confirmed by Professor Niamh Nic Daéid, director of the Leverhulme Research Centre for Forensic Science at Dundee University, in her presentation on 19 June 2018 to the hearing at the ongoing Grenfell Tower Public Inquiry. Grenfell Tower Inquiry, Hearings, Niamh Nic Daéid, 19 June 2018, https://www.grenfelltowerinquiry.org.uk/hearings, accessed 31 July 2018.
71. Grenfell Tower Inquiry, Hearings, Daniel Brown, 4 June 2018, https://www.grenfelltowerinquiry.org.uk/hearings, accessed 31 July 2018.
72. The Grenfell Tower Public Inquiry is an independent public inquiry ordered by the UK prime minister Theresa May that aims to find out the circumstances that led to the disaster. It began hearing evidence in May 2018 and it might run until 2020.

73. Grenfell Tower Inquiry, Hearings, Niamh Nic Daéid, 19 June 2018, https://www.grenfelltower-inquiry.org.uk/evidence/professor-niamh-nic-daeids-presentation, accessed 9 February 2019.
74. The cladding of Grenfell Tower was only added during a refurbishment in 2015–16 that cost £9 million. This addition aimed to increase the energy and heating performance of the tower and to improve its external appearance.
75. Barbara Lane, 'Grenfell Tower – fire safety investigation: The fire protection measures in place on the night of the fire, and conclusions as to: the extent to which they failed to control the spread of fire and smoke; the extent to which they contributed to the speed at which the fire spread', 12 April 2018, https://assets.grenfelltowerinquiry.org.uk/documents/Dr%20Barbara%20Lane%20report%20-%20section%205%20-%207.pdf, accessed 9 February 2019.
76. BBC News, 'Grenfell Tower: What Happened', https://www.bbc.co.uk/news/uk-40301289, accessed 18 June 2018.
77. 'Grenfell Tower Inquiry: Opening Statement of the London Fire Brigade', 18 May 2018, https://www.grenfelltowerinquiry.org.uk/sites/default/files/documents/London%20Fire%20Brigade%20opening%20statement.pdf, accessed 3 August 2018.
78. The author Frances Kennett considers the Joelma fire as one of the greatest disasters of the twentieth century. This accident started as a result of a faulty air-conditioner and managed to kill nearly 200 people, creating scenes that remind one of the terrorist attacks at the World Trade Center in New York in 2001. The Joelma accident, which brought about immediate changes in Brazilian building legislation, highlights common problems and mistakes associated with contemporary urban fires, which can be avoided if relevant legislation is enforced. See Frances Kennett, *The Greatest Disasters of the 20th Century* (London: Marshall Cavendish, 1975).
79. David Cohen, 'Leaked Grenfell Dossier Reveals How Disastrous Refurbishment Turned Tower into a "Tinderbox"', *Evening Standard*, 16 April 2018, https://www.standard.co.uk/news/london/shock-grenfell-dossier-reveals-disastrous-refurbishment-turned-tower-into-a-tinderbox-a3814866.html, accessed 6 August 2018.
80. Cohen, 'Leaked Grenfell Dossier Reveals how Disastrous Refurbishment Turned Tower into a "Tinderbox"'.
81. Andrew Arden, 'Editorial: Grenfell Tower: The Greatest Failure', *Journal of Housing Law*, 20: 5 (2017), 98.
82. Ivar Holm, *Ideas and Beliefs in Architecture and Industrial Design: How Attitudes, Orientations, and Underlying Assumptions Shape the Built Environment* (Oslo: School of Architecture and Design, 2006), 333.
83. K. G. Whittick, *The Party Wall: A Short History* (Godalming: Faculty of Party Wall Surveyors, 2007), 9–10.
84. Holm, *Ideas and Beliefs in Architecture and Industrial Design*, 333.
85. Some sources suggest that the change to stone and brick as building materials was also a consequence of the limited timber resources of that period. As nine-tenths of timber was used to fuel iron and glass furnaces, the oak left for the use of the navy was insufficient.
86. Bare, *Fundamentals of Fire Prevention*, 123.
87. Barry D. Yatt, *Cracking the Codes: An Architect's Guide to Building Regulations* (New York: Wiley, 1998), 37.
88. Barry D. Yatt, *Cracking the Codes: An Architect's Guide to Building Regulations* (Chichester: John Wiley & Sons, 1998), 11.
89. The Secretary of State, by Act of Parliament, 'is responsible for ensuring that the health, welfare and convenience of persons living in or working in (or nearby) buildings is secured. This Act is called Building Act 1984 and one of its prime purposes is to assist in the conservation of fuel and power, to prevent waste, undue consumption, misuse and contamination of water. It imposes on owners and occupiers of buildings a set of requirements concerning the design and construction of buildings and the provision of services, fittings and equipment used in (or in connection with) buildings.' Ray Tricker and Rozz Algar, *Building Regulations in Brief, Fifth Edition* (Oxford: Butterworth-Heinemann, 2007), 1.
90. Whittick, *The Party Wall*, 9.
91. Sam Webb, 'Councils Should Learn from King's Cross Fire', *The Architects' Journal*, 230 (May 2009), 16.
92. In the fact sheet covering the tragic accident at King's Cross issued by the London Fire Brigade, Richard L. Best writes that the 'cause has yet to be determined, but the initial investigations have ruled out various possibilities such as arson, an electrical fault, burning fluff and grease below the escalator in the machine room, or a build-up of gas such as methane in the area at

the top of the escalator. Only two causes cannot yet be discounted: Smoking materials and friction in the escalator external machinery parts. Of the two possibilities, the first is considered to be the most likely contender and as a result a complete ban on smoking on London's Underground system, has been introduced. The annual inspection of the station was carried out in April this year. The brigade's report said sacks of rubbish and cables were found on platforms, buckets of waste oil in tool rooms, combustible materials in a storeroom, and the ventilation plant room was in a dirty state.' 'King's Cross Fire, 18 November 1987', *The Times*, 19 November 1987.

93. Webb, 'Councils Should Learn from King's Cross Fire', 16.
94. The word comes from the ancient Greek mythical god Pan, and was originally attributed to the mass emotion that spread amongst the Persian army during the Battle of Marathon against the Greek army.
95. S. D. Sime, 'The Concept of Panic' in *Fires and Human Behaviour,* ed. David V Canter (London: Fulton, 1990), 63–4.
96. Blackstone and Fire Protection Association, *A History of the British Fire Service*, 252–3.
97. Sime, 'The Concept of Panic', 68–9.
98. Its name in Greek means *inconsumable* or *inextinguishable*, and it has been used since ancient times.
99. A. M. Low, *Romance of Fire* (Worcester and London: Trinity Press, 1941), 205.
100. The theatre fire curtain was invented by Captain Sir Eyre Massey Shaw after a series of immense fires that took place in the 1880s. Following a big fire at the Alhambra Theatre in Leicester Square in London, during which the Prince of Wales was nearly killed, Shaw published an article called 'Fires in Theatres' that called on every theatre to make use of fire curtains made of metal, which, in case of a fire, could separate the theatre from the auditorium.
101. Low, *Romance of Fire*, 208.
102. Whittick, *The Party Wall*, 19.
103. Lyons, *Fire*, 88–9.
104. *Symposium on Fire Test Methods, Presented at the Fourth Pacific Area National Meeting* (Los Angeles: American Society for Testing and Materials, 1963), 3–5.
105. J. Randall Lawson, *NIST Technical Note 1628: A History of Fire Testing* (Gaithersburg, MD: National Institute of Standards and Technology, 2009), 21–2.
106. Lyons, *Fire*, 94.
107. Low, *Romance of Fire*, 212–13.
108. R. Y. Qassim, 'A Program for Computational Fire Dynamics' in *Fire Safety Engineering*, ed. D. N. Smith (Bedford: Fluid Engineering Centre, 1989), 87.
109. Lyons, *Fire*, 96–7.
110. Qassim, 'A Program for Computational Fire Dynamics', 94.
111. Qassim, 'A Program for Computational Fire Dynamics', 97.
112. Bruno Latour, *Science in Action: How to Follow Scientists and Engineers Through Society* (Cambridge, MA: Harvard University Press, 1987), 67.
113. Latour, *Science in Action*, 69.
114. Gaston Bachelard, *The Dialectic of Duration*, translated by Mary McAllester Jones (Manchester: Clinamen Press, 2000), 34–5.
115. Cristina Chimisso, *Gaston Bachelard: Critic of Science and the Imagination* (London and New York: Routledge, 2001), 72.
116. Bachelard, *The Dialectic of Duration*, 20.
117. Gaston Bachelard, *The Poetics of Space*, translated from the French by Maria Jolas, with a new foreword by John R. Stilgoe (Boston: Beacon Press, 1994), 9.
118. Henri Bergson, *Creative Evolution*, translated by A. Mitchell (New York: Henry Holt, 1911), 11.
119. Bergson, *Creative Evolution*, 86.
120. Bergson, *Creative Evolution*, 81.
121. Bergson, *Creative Evolution*, 82.
122. Bergson, *Creative Evolution*, 82.
123. Bergson, *Creative Evolution*, 82.
124. Bergson, *Creative Evolution*, 175.

7
Conclusion

> I can no longer understand how we could have overlooked the ubiquity of non-erotic aggression and destruction, neglecting to give it its due place in the interpretation of life.
>
> Sigmund Freud, *Civilisation and its Discontents* (1930)

The concept of memory has undergone significant alteration in recent times. There is a great deal of confusion and uncertainty about what is worthy of preservation. Answering the question, 'what should we remember?' is difficult and fraught with complexity. The current desire to record and preserve memories from the past seems unparalleled and without limit. It manifests on both an individual and collective level. As we saw in chapter 1, Pierre Nora attributes this phenomenon to a contemporary anxiety about the future and its uncertain development, and calls it *the duty to remember.* For the sociologists Ulrich Beck and Anthony Giddens, the future is similarly uncertain because we live in a *risk society*, which is characterised by an ongoing anticipation of the future and fear of irreversible destruction, which in turn creates the need to invest in the stockpiling of memory.

The main focus of this book has been architectural conservation, a discipline that is increasingly receiving great attention from governments, professionals and the general public. This interest can be measured by the continuous reorganisation of legislation as well as growing academic engagement. One of the key points of concern in contemporary conservation discourse is the lack of a consistent policy. This means, for example, that theoretical dilemmas arise when trying to establish a rationale for the conservation of post-war architecture. Often buildings from that period do not fall within any conservation policy. Furthermore, conservation policies are based on Western values and ideas, and deal only with the exceptional and the unique. I have also argued that one of the most critical problems of conservation is inherent in its nature. This is its destructive effect on the linear evolution of time in buildings, which brings about what the architect Rem Koolhaas has called *Cronocaos* (see chapter 1). On the one hand, restoration reverses time so that a building returns to previous orders, as it once stood, and on the other, conservation freezes time so that a building's current state is extended into the future.

The problems associated with conservation form the foundations of my research in this book, and also set its objective, which was to establish an alternative conceptual approach to architectural conservation. This took place by navigating across various disciplines such as archival theory, psychoanalysis, classical mythology, conservation history and theory, and architectural history and theory. Reflections on the topic also gave rise to the invention of an unusual methodology. This is the methodology of fire, an inherently ambivalent methodology that is also constructive and illuminating.

The approach to conservation was made though archival theory. This decision was based on the fact that listed buildings operate like archives in the sense that both aim at the preservation of memory. Chapter 2 therefore explored the concept of the archive and elaborated on Jacques Derrida's understanding of it. My interest in Derrida's archival theory is based on his association of the archive's function with the function of the Freudian unconscious. In both instances, this function incorporates three sequential stages: inscription, storage and retrieval. Derrida makes an additional conceptual association, which is that the theory of psychoanalysis is also a theory of the archive. Just like humans, archives can be perceived as having life and death drives. The drives of archives – particularly the death drive, or archive fever as Derrida calls it – will inform my final arguments.

The concept of individual and collective memory was then developed in line with archival theory. A long tradition in Western thought associates the memory of the individual with the function of the archive. The three-faceted operation of the archive – inscribing, storing, retrieving – also reflects the three stages of the memorising process that take place in the individual. Plato likens memory to a block of wax on which we imprint our thoughts and perceptions. In ancient Rome memorisation, which was an art, took place by inserting and storing mental images in mental places, and recollection occurred by retrieving these images through mental strolls. This tripartite function of the memory of the individual is also evident in more recent discourses, as in Freud's psychoanalytic theory.

Lastly I explored the concept of collective memory, a term coined by Maurice Halbwachs, which also appears to operate as an archive. For Halbwachs, memory has social dimensions. It depends on the social framework in which it is created and the material objects present in that framework. Recollection demands a reconstruction of this social framework in the present; thus memory can never regain its original essence. As soon as the social group to which a memory belongs dies, collective memory also dies and is replaced by history. For the social anthropologist Paul Connerton, collective memory is transferred to the present through the practice of commemorative ceremonies and bodily practices. Collective memory similarly performs archival functions in the sense that processes of inscribing, storing and retrieving still take place. While for individual memory these processes are internal, for collective memory they are external and associated with social frameworks or cultural formations and communications. With regard to individual memory, although memorisation is purely internal, it still demands an archiving process. Both ancient and modern theories often suggest that memory needs to be externalised onto some sort of substrate so that it can later be used for recollection. Therefore the function of individual memory can be considered as external as the function of collective memory, which demands objects or practices to refer to. Through looking at concepts relevant to archival theory, both individual and collective memory appear to operate with an archival logic. Effectively the distinction between the two types becomes superfluous.

Chapter 3 focused on methodology. This book dealt with a specific memory in architecture, the memory of fire. It explored how that memory took shape in the form architecture took, as well as the forms it left behind. The choice of fire is relevant in that the primitive hut was erected around it: therefore it is architecture's first memory. At the same time, as fire is one of the major causes of architectural destruction, fire is the *last* memory of much architecture. The study of the memory of fire pointed towards the deployment of specific philosophical theories. This was the work of Gaston Bachelard, who is known for his views on memory and fragmented time, and also renowned for his psychoanalytic approach to fire. Taking into consideration fire's conflicting nature, an opposing theory from a philosophical rival was employed. This was the work of Henri Bergson, who propagated the concept of temporal continuity. It is not accidental that all arguments throughout this book were based on the theoretical conflict between these two philosophers, for this follows Bachelard's own views on the advancement of knowledge. Bachelard explains that the formation of objective knowledge can only be achieved through modern sciences. It is a constant dialectical process that is often aggressive and polemical, as it can potentially shake the foundations of a whole discipline. This is what he calls an epistemological rupture. Science can refuse its own past and reorganise itself radically from scratch.

Bachelard himself uses this model of knowledge production in his own philosophy and encourages its employment. This served as an invitation to set his philosophy against that of Bergson. For Bachelard, in the core of scientific knowledge there is a dream, and he speculates that the first thing that the human mind must have dreamt of is fire. Consequently the starting point of science, which is the essence of scientific knowledge, is fire. The theoretical conflict between the opposing theories of Bachelard and Bergson carries a memory of the element too.

To support the omnipresence of fire in knowledge and conflict, I referred to Jean-Pierre Vernant's interpretation of the Promethean myth and the ancient philosophy of Heraclitus. According to Vernant, fire carries the memory of the divine conflict between Prometheus and Zeus, while the possession of the element suggests a *technics*, a technical ability, which is the link between fire and knowledge. In Heraclitus' philosophy, the cosmos emerged through the divine interaction of the four elements: fire, earth, water and air, with fire as the leading force. It is to be found in flux, and this movement is the result of a universal conflict. Conflict is above all necessary as it leads to the creation of life and, just as for Bachelard, it is associated with the eternal nature of fire. The intention to draw parallels between fire, conflict and knowledge was that the proposed methodology of conflict became a methodology of fire.

The second part of this chapter looked closely at our current knowledge of fire, which appears to be muddled. It is a mix of personal intuition and scientific knowledge, and it is always taken for granted. For this reason I looked into some of its multiple manifestations throughout the years, in particular those appearances that speak of this confused nature. Drawing mainly on the extensive and pioneering work of the social anthropologist J. G. Frazer, I offered a retrospective inquiry into fire's manifestation in mythology, ancient rituals and festivals. I referred to the advancements that took place during the eighteenth century in the studies of combustion, which according to Bachelard can be considered as an epistemological rupture, and which brought about a radical advancement in our knowledge of the element. The archival perspective clarifies why contemporary, scientific knowledge of fire is merged with ambiguous notions and fallacies from the past that stretch back to ancient times. If ancient and current knowledge are perceived as two distinct archives, the replacement of one by another naturally provokes a degree of confusion.

Chapter 4 reflected on the obvious association of architectural heritage with the notion of the archive. After elaborating on the two opposing currents in modern conservation history – Viollet-Le-Duc's restoration movement in France and John Ruskin's conservation movement in England – I focused on current issues in conservation. The contemporary approach to the theory and practice of architectural conservation, which is a phenomenon rooted in Western ideology, is based on the attempt to unite both restoration and conservation principles into a single strategy, and is directed and regulated by international charters and agreements. My argument focused on English conservation practice, which has the legal power to grant permission for the periodic renewal of buildings' elements, repairs that do not affect the current state and look of buildings, interventions that might increase knowledge of the past, and restoration processes that can offer buildings more heritage value. This set of interventional processes has an unavoidable impact on the linear evolution of time on buildings.

The impact of conservation practice on the linear evolution of time affects the function of memory in architecture, which was initially discussed through Bachelard's theory of time and memory. According to his theory of repose, time is discontinuous in the sense that it can be broken down to at least two instants, so that it has a dialectic. Every duration and thus also every evolution starts out of nothingness and, as it is discontinuous, can be broken down to instants. In order to experience an evolution, one needs to activate the rhythm of its constitutive discontinuity, of creation and destruction, otherwise evolution cannot take place. Memory for Bachelard is also discontinuous and defined by consecutive breaks. Influenced by Halbwachs, Bachelard claims that memory depends upon the social framework in which it is created: thus the more secure the framework is, the sounder the memory. For recollection to take place, one needs to remember these social frameworks, which mark the past as decisive moments, and this can occur through the employment of the imagination. English conservation practice abides by Bachelard's notion of time and memory in the sense that both conservation and restoration processes punctuate the evolution of a building at specific points in time, either by freezing time in the present or by reversing it to a previous state. Accordingly, listed buildings can also be perceived as frameworks in which memories are fixed and secured. If the framework is successfully preserved or restored, memory can be cued and recollection can successfully take place.

English conservation practice was then approached through the opposing theory of Henri Bergson, according to which archives, and also listed buildings, can be perceived as sites of forgetting. Following Bergson, there are two types of time: a homogeneous one, the time of physics that can be measured in space, and real time or duration, which is heterogeneous and can only be experienced with the method of intuition, a method Bergson formulated for qualitative divisions. With regard to memory, Bergson explains that there are two main types. One is habit memory, which deals with the memory that is acquired through the repetition of an action, as if it were a habit, and the other is representational memory, which records every detail of our life in the form of memory-images. In order to produce a memory-image in the present, the living body, which is the centre of action that is called the body-image, needs to trace it amongst the totality of all memory-images present in memory and bring it forth into the present. Recollection is therefore the actualisation of a memory-image from a virtual place or substratum. This imaginary movement of bringing a memory out of some sort of storage place gives the wrong impression, as if it comes out of an archive. There is also a third type of memory in Bergson's philosophy, which is called pure memory, and refers to the totality of memory-images found in a virtual state. In the process of recollection, the filtering out that takes place when a memory-image from a virtual state is actualised can in fact be considered as a process of forgetting.

Bergson's memory is not archival in a conventional sense. Memorisation does not involve imprinting on any substratum such as a wax tablet or the unconscious. The emergence of the archive occurs only in the process of recall, when memory is actualised. In this sense the totality of memory does not reside in archives but elsewhere, and therefore whatever is to be found in archives is merely a reduction of memory. This argument is contextualised within conservation practice. Conservation practice is the body-image, the centre of action, which filters out the totality of memory-images that makes up the complete past of a building by preserving only an aspect of it. Therefore listing, which is in essence a process of archivisation, preserves only a fraction of the totality of memory and thus listed buildings can be considered sites of forgetting.

Chapters 5 and 6 took a step back and looked at the broader relationship between architecture and memory by focusing on the study of a specific element, namely fire. Chapter 5 looked into architectural evolution from the ancient flames of the hearth to contemporary architecture. This was approached through two perspectives. One dealt with the way architecture adapts itself to accommodate fire based on the thermal comfort it provides, and the other considered fire exclusively as energy, exploring its proliferation in architectural space in energetic terms. With regard to architecture's adaptability according to thermal comfort, I discussed how the fire of the hearth once occupied a central position in architectural space and functioned as a gathering point. During the first years of modernity, there was a shift. The centrality of the hearth was gradually replaced by fireplaces, radiators and underfloor heating. People found themselves surrounded by multiple centres from which fire is now dislocated. This dislocation becomes final in sustainable design where the combustion of energy is replaced or done without. The spaces fire once inhabited are redundant.

An energy-centred investigation of architectural evolution, by contrast, perceives a building as a composite of different energies. The energy that fire produced within the primitive hut was initially expended for maintenance but was later used for the construction of buildings too. Both perspectives were explored to understand how architecture records memory. Following Bachelard's philosophy, I demonstrated that architecture is perceived as a porous entity that reflects the needs and demands of society. In archival terms, architecture operates as an archive in which a reduced memory of its entire past is recorded and stored. This memory can be used as a guideline for future architectural creations, while the memory found within the architecture/archive is externalised each time a new building is built. I then juxtaposed Bachelard's approach to evolution with that of Bergson. For Bergson, evolution is a dynamic process in constant flux, which is driven by *élan vital*. This vital force or impulse keeps evolution in ceaseless flow. The problem of understanding evolution, Bergson explains, is based on a more general inability of the intellect to experience the flow of duration and thus also the duration of evolution. While duration can be experienced instinctively, we have the tendency to perceive it with our intellect, which breaks it down to instants. These instants do not represent the real essence of duration but are merely snapshots of its continuity. Equally, if architectural evolution is perceived as a process that emerges in real time, in duration, then the entirety of its past exists at any time in the present, with both past and present pointing towards the future. For Bergson, therefore, architecture always carries with it the totality of memory of its entire past.

In chapter 6 I looked into a key moment of architectural evolution, which is architecture's temporal death caused by fire, and investigated the way architecture absorbs this event. Fire in buildings is always imminent, and when it breaks out, it is often unpredictable. As a result, societies respond in multiple ways to tackle this threat. Today buildings are protected with active and passive fire protection systems. There are organised fire-fighting forces, appropriate building legislation and a growing investment in fire prevention research. Architecture's response to fire accidents was first discussed through Bachelard's philosophical perspective. Architecture's temporal death caused by fire punctuates its overall evolution. A destroyed building is useful knowledge that can be used for the advancement of fire-related research, building legislation and so on. The moment a building is destroyed marks a period during which architectural knowledge is reorganised. What is successful and efficient will be repeated. The rest belongs to the past.

Architecture's response to accident was then developed according to Bergson's theory of evolution. For Bergson, evolution is an open-ended, dynamic process in flux that develops in duration, and the temporal death of architecture is merely an aspect of this evolution. Bergsonian evolution does not only refer to organic changes that take place in our physical nature but also includes advancements in our intellectual and material culture. Yet if evolution is perceived in Bergsonian terms, in duration, then it carries with it the totality of its past, including every advancement in intellectual and material culture. Both chapters 5 and 6 argued that architecture has clear archival properties and concluded that every building carries with it either a reduced or a complete memory of architecture's entire past. Contextualising this within the main argument of this book, the fact that architecture stores either a totality or a contraction of its entire past calls into question the need for conservation, and proclaims it essentially redundant.

A psychoanalytic approach to conservation

This book opened with a reference to Louis Benassi's *Black Umbrella*, a film triptych that explores key themes developed in my research. The 16mm reels of the triptych are archival material belonging to the London Fire Brigade. They were accidentally found in a disused fire station in London and depict flaming buildings. Archives, buildings and fire are the three main topics of this research, but up to this point they have been examined either in isolation or in combination of two at a time: buildings and archives, fire and archives, and buildings and fire. Inspired by Benassi's triptych, where archives, buildings and fire are powerfully intermingled, I propose here a new approach to architectural conservation.

In *Black Umbrella* Benassi appropriated the use of the archive to his artistic needs. Whereas the original use of the fire brigade's archive material was the documentation of fire incidents, through a skilled editing process and an appropriate projection technique *Black Umbrella* 'is [now] shielding "our" young woman from the explosive, life threatening splinters produced by the flying bomb, the flying bomb, which incidentally could be seen as a metaphor for the heartless architects of displacement'.[1] Doing so, the filmmaker makes a bold statement about not only the limits but also the possibilities of archives. The postmodern concept of the 'death of the author' is clearly manifest, as the author's initial intention, which is the documentation of fire incidents, is absent from the triptych. This further confirms the fact that archives do not reflect a single, unified truth but a series of many valid truths that are dependent on the interpretation and/or the ends of the researcher. Hence my intention to refer to *Black Umbrella*, as it clearly celebrates the versatile, polysemic nature of the archive that also allows for a psychoanalytic interpretation. It agrees, moreover, with Derrida's overall ambition outlined in *Archive Fever* to open up 'a project of a general archiviology … [of] a general and interdisciplinary science of the archive'.[2] In this specific project of archiviology, conservation is investigated through two clear archival associations. It is first argued that both fire and buildings operate as archives. Based on these grounds and applying Derrida's theory of the archive, a further conceptual leap is possible where both fire and buildings can be examined through psychoanalysis. Fire's conflicting nature has the capacity to create life and at the same time consume it in its flames: it is a pure manifestation of the drives. Through the psychoanalysis of fire therefore it is confirmed that the element carries both life and death drives, which is what Bachelard calls the Empedocles complex.

Benassi's triptych portrays the burning down of buildings. Superimposed on it is the projection of a female figure carrying a black umbrella that protects her from exploding building elements. *Black Umbrella* depicts the imminence of architecture's destruction as well as the urgency of safeguarding it. If it is accepted that architecture operates as an archive, it can be approached psychoanalytically through the theory of the drives. On the one hand the architecture-archive reflects the life drive, which is the drive that leads towards the creation and preservation of life and refers to every built structure around us. On the other hand, as architecture is an archive, it also carries an intrinsic quality of archive fever. Archive fever, which relates to the Freudian death drive, pushes towards extinction, towards an inorganic state. It reflects the desire to return to the absolute beginning of architecture, which is the element of fire. In this sense, architecture preserves memory through an alternating process of creation and destruction.

Translating this into concrete terms, one simply needs to reflect on any building. Buildings stockpile a great deal of memory that stretches back to the beginning of architecture, to the primitive hut. This memory is stored in their form, in the typology and materials, and in the technology that is used to construct, power and maintain them. Buildings are assembled in compliance with relevant legislation, which is generated by the success or failure of previous buildings. Legislation itself is arguably a type of memory that is externalised onto the form of architecture. Buildings are also catalysts for the existence and prolonging of collective memory. Collective memory does not only depend on the social framework but also on the material space in which it is created. If the architecture of a city changes drastically, this will have an inevitable destructive effect on collective memory. Even though memory is stored in architecture, its longevity is fragile. Buildings cannot escape the impact of time. They unavoidably decay and degenerate, and thus the memory they hold slowly fades. To avoid the complete eradication of memory, new buildings must be built, creating a vicious circle of creation and destruction that leads to the preservation of memory. This schema is analogous to the myth of memory according to which Mnemosyne's enactment on behalf of remembering takes place through the process of construction, yet it is always counteracted and gradually overshadowed by the opposing work of Lesmosyne on behalf of forgetting. Therefore, to preserve memory is to sustain the cycle that articulates in a simultaneous occurrence the complete dimension of the mythological entity of Memory, of both Mnemosyne and Lesmosyne.

An example of an architecture that demonstrates this alternating play between remembering and forgetting as a means for the preservation of memory is the work of Giovanni Battista Piranesi (1720–1778). The famous Italian topographical engraver, who is rarely recognised for his work as a practising architect,[3] puts particular emphasis on the emotive and creative power of ruins. His work was developed partly as a reaction to a new interest in the achievements of the ancient Greeks that emerged during the eighteenth century, which undermined the superiority of the Romans. In the 'finest and most influential of all his polemical-archaeological works',[4] *Il Campo Marzio dell'Antica Roma* (1762), a folio of six etchings depicting particular views or *vedute* of Rome, Piranesi demonstrates visions of the city's past based on his extensive archaeological and architectural knowledge and enhanced by his imaginative interventions. In his *vedute* of Rome, Piranesi depicts monuments in ruins, fragmented, while the surrounding landscapes are presented with a great attention to detail, as they stood in mid-eighteenth-century Rome. The monuments witness the passing of time as imagined by the artist, and the rest of Rome's topography stays intact without undergoing any transformation. Commenting on Piranesi's selected erasure of history, the art historian John Pinto points out that the 'organic process of decay, together with cannibalistic recycling of ancient materials into modern edifices, inevitably revealed the underlying structural features of Roman buildings, making visible aspects of these monuments that were never intended to be seen'.[5] Piranesi's *Campo Marzio* reflects a deliberate interplay between architectural remembering and forgetting. The intention of the artist through this selective fragmentation of the urban landscape is to produce a memory in which the supremacy of Rome exceeds that of ancient Greece. In the context of the main argument, *Campo Marzio* can also be perceived as a powerful illustration of both life and death drives, through the interaction of which memory in architecture can be sustained in the present and prolonged into the future. *Campo Marzio* therefore has the same function as *Black Umbrella*. They both reflect the life and death drives of architecture, and they are both pivotal for the preservation of memory.

The interpretation of architecture in accordance with the theory of the drives allows for a further conceptual association. Architectural conservation, which is in essence an official process of archivisation, aims to protect the material aspect of architecture. It seeks to prolong the building's integrity by either freezing it in time or returning it to previous orders. Through a psychoanalytic perspective, conservation emerges as a discipline that on the one hand protects the life drive and on the other represses the death drive, or archive fever, thus preventing full architectural expression. It prevents the decay and degeneration of buildings and therefore interferes with the way memory is innately preserved and prolonged in architecture.

Considering architectural conservation as a repression of the death drive allows for the resolution of key conceptual problems or inconsistencies inherent in conservation's nature. It was mentioned earlier, for instance, that post-war architecture does not usually fall under any regime of conservation. A psychoanalytic interpretation of this argument suggests that the death drive is in this case freed from repression. The desire to destroy post-war buildings relates to the desire to obliterate the memory or rather the trauma associated with them, so that new buildings can replace them. One can suggest further that the architecture of the modern movement was essentially an architecture of the death drive, in the sense that it aimed to 'erase the traces'[6] and 'endure less than us'.[7] Therefore if we are to preserve the true spirit of modernist architecture, the death drive should be released so that its complete memory is preserved through its material absence. This statement agrees in principle with Owen Hatherley, who sees the conservation of modernism as the absolute conspiracy 'with the people who have always opposed it'.[8] This desire to obliterate the memory of buildings is reminiscent of the ongoing series of installation projects titled *City on Fire* by Thyra Hilden and Pio Diaz that has taken place since 2005 in various cities around Europe. By projecting fire images on monuments or other buildings of cultural significance, the two artists aim to 'reveal the fragile and transitory nature of these man-made constructions, and thereby destabilise prevailing order'.[9] My interpretation of these installations is that through temporary events of symbolic destruction they achieve a conceptual articulation of the death drive that questions the value and significance of the roots of Western civilisation.

Returning to the proposal of approaching conservation through the drives theory, the fact that conservation policies apply only to buildings that are exceptional and unique suggests that the mediocre and ordinary are beyond appreciation. This reflects a release of the death drive. The prospect of establishing a common international conservation agenda is also problematic, as criteria for releasing the death drive differ amongst cultures. Therefore the proposed theory of conservation not only responds to the historical value of buildings but also embraces an additional parameter: it considers the existence of the drives and promotes their externalisation.

Following psychoanalytic theory, the drives must be in *peaceful balance*. As discussed in chapter 2, in Freudian psychoanalysis the death drive has a twofold manifestation: one that is external and expressed as a sadistic aggression and one that is purely internal and expressed as masochism. Freud claims that both manifestations are normal, and in fact are necessary and beneficial for the organism. Evilness and aggression are indeed parts of our nature, and they have to be externalised. In *Civilisation and its Discontents*, Freud attributes the glory of civilisation to the simultaneous workings of the life and death drives, where the death drive is diverted or sublimated. He explains that civilisation serves the life drive (Eros), which brings people together and makes them into a solid, functional unit. He acknowledges that among people there is also innate aggression, which reflects the working of the death drive (Thanatos).[10] Later, however, he adds that

> the development of civilisation is no longer obscure to us. It must show us the struggle between Eros and Death, between the life drive and destructive drive, as it plays out in the human species. This struggle is the essential content of all life, and therefore the development of civilisation can simply be characterised as the struggle for life of the human species.[11]

As the death drive is innate but essential to life, it is imperative that the evil and aggression with which it is expressed are sublimated, channelled creatively as a *will to power*. This is possible by abolishing the dualistic model of the life/death drive, according to which the death drive counteracts the righteousness of the life drive, and instead seeing the two as *one* fundamental positive force of our nature.[12] With regard to conservation this psychoanalytic approach accepts that progress in architecture is achieved through the *equal* workings, or struggle, of the life and the death drives, where destruction is not necessarily considered as something negative. After all, as Freud points out, our civilisation would not develop unless it was constantly threatened by destruction. The same applies to architecture.

A powerful illustration of an artistic attempt to *balance* the architectural life and death drives through a sublimation of the latter is the temporary creation of Skoghall Konsthall that took place in 2000 by the Chilean artist Alfredo Jaar. Jaar was invited to the Swedish town of Skoghall, which houses the workforce of a large pulp mill that operates in the area, to propose a rejuvenating work of public art. Skoghall was a 'dormitory town, with no civic space for the exhibition of art, no amenities whatsoever for the production of culture of any sort'.[13] In response, Jaar proposed to design and build a community art space, Skoghall Konsthall, commissioned by the pulp mill company, in which local artists could exhibit their work (fig. 7.1a). The only difference from any other exhibition space was the fact that 24 hours after the opening, Skoghall Konsthall was to be burnt to the ground in the presence of the local community (fig. 7.1b).

A psychoanalytic reading of this project suggests that the creation and almost immediate destruction of Skoghall Konsthall is a successful and vibrant articulation of both life and death drives, where destruction is welcomed and in fact celebrated as a healthy and beneficial option for the progress of civilisation. It therefore stresses how through a psychoanalytic reading, controlled destruction must be an integral part of a healthy conservation agenda.

The theoretical framework developed in this book is entirely conceptual. The conclusion does not define an end, but rather a beginning: one that is open to interpretation, such as further theoretical work, diverse artistic interventions or even actual praxes of destruction. The interdisciplinary approach employed in this book brings to light a relationship with potential future and dynamism. This is the relationship between architectural conservation and psychoanalysis. Exploring the possibilities and crossing the boundaries of these two disciplines can lead to something new and radical. Lastly, looking at architecture though the drives theory creates a disciplinary bridging. One of the effects of

Figure 7.1a, b The opening and closing of Alfredo Jaar's Skoghall Konsthall, Sweden, 2000. © Alfredo Jaar.

modernism was the division of the urban landscape into new and old, which encouraged the creation of two opposing disciplines, that of the architect and that of the conservationist. Through a theory of the drives, the distinction between these two disciplines becomes superfluous. The architect is a conservationist and the conservationist is an architect.

Notes

1. Close-Up, '1 February 2011: Abandoned Archives and Forgotten Histories Remembered'. https://www.closeupfilmcentre.com/film_programmes/2011/abandoned-archives-and-forgotten-histories-remembered/, accessed 11 February 2019.
2. Jacques Derrida, *Archive Fever: A Freudian Impression*, translated by Eric Prenowitz (Chicago: University of Chicago Press, 1995), 34.
3. John Wilton-Ely, *Piranesi as Architect and Designer* (New York and New Haven: Pierpont Morgan Library/Yale University Press, 1993), i.
4. Wilton-Ely, *Piranesi as Architect and Designer*, 39.
5. John A. Pinto, *Speaking Ruins: Piranesi, Architects, and Antiquity in Eighteenth-Century Rome* (Ann Arbor, MI: University of Michigan Press, 2012), 111–12.
6. Owen Hatherley, *Militant Modernism* (Winchester and Washington, DC: O Books, 2008), 3.
7. Antonio Sant'Elia, 'Manifesto of Futurist Architecture' [1914] in *Rethinking Technology: A Reader in Architectural Theory*, eds, William W. Braham, Jonathan A. Hale and John Stanislav Sadar (London and New York: Routledge, 2007), 21.
8. Hatherley, *Militant Modernism*, 5.
9. Lisbeth Bonde, 'Video Pyromaniacs and Pictures of Fire Related Works', https://piodiaz.wordpress.com/2007/02/20/video-pyromaniacs/, accessed 26 August 2018.
10. Sigmund Freud, *Beyond the Pleasure Principle*, edited by Dufrense Todd, translated by Gregory C. Richter (Peterborough, ON: Broadview Press, 2011), 134–5.
11. Freud, *Beyond the Pleasure Principle*, 135.
12. Rob Weatherill, ed., *The Death Drive: New Life for a Dead Subject?* (London: Rebus Press, 1999), 36.
13. Michael Corris, 'White Out', *Art Monthly*, 260 (October 2002), 7.

Bibliography

Books and book chapters

Adorno, Theodor W. *Prisms*. Cambridge, MA: MIT Press, 1967.

Aqua Group. *Fire and Building: A Guide for the Design Team*. London: Granada, 1984.

Aristotle. *On the Soul, and other Treatises of Aristotle (The Works of Aristotle VI)*, translated by Thomas Taylor. Frome: Prometheus Trust, 2003.

Asenjo, F. G. *In-Between: An Essay on Categories*. Washington, DC: Center for Advanced Research in Phenomenology and University Press of America, 1988.

Ashby, Mike, and Kara Johnson. *Materials and Design: The Art and Science of Material Selection in Product Design*. Oxford: Butterworth and Heinemann, 2002.

Axelos, Kostas. *O Herakleitos kai i Filosofia*. Athens: Exantas, 1976.

Bachelard, Gaston. *The Dialectic of Duration*, translated by Mary McAllester Jones. Manchester: Clinamen Press, 2000.

Bachelard, Gaston. *The Flame of a Candle*, translated by Joni Caldwell. Dallas, TX: Dallas Institute Publications, 1988.

Bachelard, Gaston. *The New Scientific Spirit*, translated by Arthur Goldhammer. Boston: Beacon Press, 1934.

Bachelard, Gaston. *The Philosophy of No: A Philosophy of the New Scientific Mind,* translated by G. C. Waterston. New York: Orion Press, 1968.

Bachelard, Gaston. *The Poetics of Reverie: Childhood, Language, and the Cosmos*, translated by Daniel Russell. Boston: Beacon Press, 1960.

Bachelard, Gaston. *The Poetics of Space*, translated by Maria Jolas, with foreword by John R. Stilgoe. Boston: Beacon Press, 1994.

Bachelard, Gaston. *The Psychoanalysis of Fire*, translated by Alan C. M. Ross with preface by Northrop Frye. Boston: Beacon Press, 1968.

Bachelard, Gaston. *Water and Dreams: An Essay on the Imagination of Matter*, translated by Edith Farrell. Dallas, TX: Pegasus Foundation, 1983.

Ball, Philli. *The Elements: A Very Short Introduction*. Oxford: Oxford University Press, 2002.

Bare, William K. *Fundamentals of Fire Prevention*. New York: Wiley, 1977.

Bastea, Eleni, ed. *Memory and Architecture*. Albuquerque, NM: University of New Mexico Press, 2004.

Bataille, Georges. *The Accursed Share*, vol. 1, translated by Robert Hurley. New York: Zone, 2002.

Beaver, Patrick. *The Match Makers*. London: Henry Melland, 1985.

Beck, Ulrich. *Risk Society: Towards a New Modernity*. London: Sage, 1992.

Beckmann, Johann. *A History of Inventions and Discoveries*, vol. 1, translated by William Johnston. London: [s. n.], 1817.

Bentham, Jeremy. *Auto-Icon; or, Farther Uses of the Dead to the Living: A Fragment*. London: privately published, 1842.

Bergson, Henri. *An Introduction to Metaphysics*, translated by T. E. Hulme. Indianapolis: Hackett, 1999.

Bergson, Henri. *Creative Evolution*, translated by A. Mitchell. New York: Henry Holt, 1911.

Bergson, Henri. *Duration and Simultaneity: Bergson and the Einsteinian Universe*, translated by Leon Jacobson. Manchester: Clinamen Press, 1999.

Bergson, Henri. *Matter and Memory*, translated by M. N. Paul and S. Palmer. New York: Dover, 2004.

Bergson, Henri. *Time and Free Will*, translated by F. L. Pogson. London: Elibron Classics, 2005.

Betsky, Aaron, and Erik Adigand. *Architecture Must Burn: A Manifesto for an Architecture Beyond Building*. London: Thames & Hudson, 2000.

Bevan, Robert. *The Destruction of Memory: Architecture at War*. London: Reaktion, 2006.

Bird, Eric L., and Stanley J. Docking. *Fire in Buildings*. London: A & C Black, 1949.

Blackstone, G. V. and Fire Protection Association. *A History of the British Fire Service*. Moreton in Marsh: Fire Protection Association, 1996.

Borden, Iain et al., eds, *The Unknown City: Contesting Architecture and Social Space*. Cambridge, MA and London: MIT Press, 2001.

Borges, Jorge Luis. *The Aleph, Including the Prose Fictions from the Maker*, translated by Andrew Hurley. London: Penguin, 2000.

Borgmann, Albert. *Technology and the Character of Contemporary Life*. Chicago and London: University of Chicago Press, 1984.

Burgess-Wise, David. *Fire Engines and Fire-Fighting*. London: Octopus, 1977.

Calvino, Italo. *Invisible Cities*, translated by William Weaver. London: Vintage, 1997.

Canter, David, ed. *Fires and Human Behaviour*. Chichester: John Wiley & Sons, 1980.

Carruthers, Mary J. *The Book of Memory: A Study of Memory in Medieval Culture*. Cambridge: Cambridge University Press, 1990.

Carruthers, Mary J. *The Craft of Thought: Meditation, Rhetoric, and the Making of Images, 400–1200*. Cambridge: Cambridge University Press, 2000.

Casey, Edward S. *Remembering: A Phenomenological Study*, second edition. Bloomington, IN: Indiana University Press, 2000.

Castino, G. T., and T. Z. Harmathy, eds, *Fire Risk Assessment*. Philadelphia: American Society for Testing and Materials, 1982.

Chamberlin, E. R. *Preserving the Past*. London: Dent, 1979.

Chambers, Sir William. *Plans, Elevations, Sections, and Perspective Views of the Gardens and Buildings at Kew in Surrey*. Farnborough: Gregg Press, 1966.

Chappell, Timothy D. J. *Reading Plato's Theaetetus*. Indianapolis: Hackett, 2005.

Chimisso, Cristina. *Gaston Bachelard: Critic of Science and the Imagination*. London and New York: Routledge, 2001.

Clarke, Bruce, and Linda Dalrymple Henderson. *From Energy to Information: Representation in Science and Technology, Art, and Literature*. Stanford, CA: Stanford University Press, 2002.

Colomina, Beatriz. *Privacy and Publicity: Modern Architecture as Mass Media*. Cambridge, MA: MIT Press, 1996.

Colquhoun, Alan. *Essays in Architectural Criticism: Modern Architecture and Historical Change*. Cambridge, MA and London: MIT Press, 1981.

Connerton, Paul. *How Societies Remember*. Cambridge: Cambridge University Press, 1989.

Consenstein, Peter. *Literary Memory, Consciousness, and the Group Oulipo*. Amsterdam and New York: Rodopi, 2002.

Cote, Arthur E. *Fundamentals of Fire Protection*. Quincy, MA: National Fire Protection Association, 2004.

Crinson, Mark, ed. *Urban Memory, History and Amnesia in the Modern City*. Oxford and New York: Routledge, 2005.

Cullinan, Nicholas, ed. *Tacita Dean: Film*. London: Tate Publishing, 2011.

Dalzell, Alexander. *The Criticism of Didactic Poetry: Essays on Lucretius, Virgil and Ovid*. Toronto: University of Toronto Press, 1996.

Deleuze, Gilles. *Bergsonism*, translated by Hugh Tomlinson and Barbara Habberjam. New York: Zone, 1988.

Derrida, Jacques. *Archive Fever: A Freudian Impression*, translated by Eric Prenowitz. Chicago: University of Chicago Press, 1995.

Derrida, Jacques. *Dissemination*, translated by Barbara Johnson. London: Athlone Press, 2004.

Derrida, Jacques. *The Post Card: From Socrates to Freud and Beyond*, translated by Alan Bass. Chicago and London: University of Chicago Press, 1987.

Detienne, Marcel, and Jean-Pierre Vernant, eds, *The Cuisine of Sacrifice Among the Greeks*, translated by Paula Wissing. Chicago and London: University of Chicago Press, 1989.

DiPippo, Ronald. *Geothermal Power Plants: Principles, Applications and Case Studies.* Oxford and Heidelberg: Elsevier, 2009.
Eggert, Paul. *Securing the Past: Conservation in Art, Architecture and Literature.* Cambridge: Cambridge University Press, 2009.
Eisenberg, David, and Peter Yost. 'Sustainability and Building Codes' in *The Sustainable Urban Development Reader*, edited by Stephen Wheeler and Timothy Beatley. London: Routledge, 2004.
Eliade, Mircea. *The Myth of the Eternal Return, Or, Cosmos and History*, translated by W. R. Trask. Princeton: Princeton University Press, 1991.
Emsley, John. *The Shocking History of Phosphorus: A Biography of the Devil's Element.* London: Macmillan, 2000.
Fernández-Galiano, Luis. *Fire and Memory: On Architecture and Energy*, translated by Gina Cariño. Cambridge, MA: MIT Press, 2000.
Fink, Eugen. *Nietzsche's Philosophy*, translated by Goetz Richter. London and New York: Continuum, 2003.
Fire Department, *Manual of Firemanship: A Survey of the Science of Fire-Fighting*, book 5, *Ladders and Appliances* (London: Her Majesty's Stationery Office, n.d.).
Fire Department, *Manual of Firemanship: A Survey of the Science of Fire-Fighting*, book 6, *Breathing Apparatus and Resuscitation* (London: Her Majesty's Stationery Office, n.d.).
Fire Department, *Manual of Firemanship: A Survey of the Science of Fire-Fighting*, book 9, *Fire Protection of Buildings* (London: Her Majesty's Stationery Office, n.d.).
Fire Department, *Manual of Firemanship: A Survey of the Science of Fire-Fighting*, book 10, *Fire Brigade Communications and Mobilising* (London: Her Majesty's Stationery Office, n.d.).
Fitch, James Marston. *Historic Preservation: Curatorial Management of the Built World.* Charlottesville: University Press of Virginia, 2001.
Fitzgerald, Robert W. *Building Fire Performance Analysis.* Hoboken, NJ: John Wiley & Sons, 2004.
Forty, Adrian, and Susanne Küchler, eds, *The Art of Forgetting.* Oxford and New York: Berg, 1999.
Forty, Adrian. *Words and Buildings: A Vocabulary of Modern Architecture.* London: Thames & Hudson, 2000.
Frazer, James George. *Myths of the Origin of Fire.* New York: Barnes and Noble, 1996 [1930].
Frazer, James George. *The Golden Bough: A Study of Magic and Religion.* Ware: Wordsworth Editions, 1993 [1890].
Freud, Sigmund. *Beyond the Pleasure Principle*, edited by Dufrense Todd, translated by Gregory C. Richter. Peterborough, ON: Broadview Press, 2011.
Freud, Sigmund. *On Metapsychology: The Theory of Psychoanalysis: 'Beyond the Pleasure Principle', 'The Ego and the Id' and Other Works*, edited by Angela Richards and translated under the general editorship of James Strachey. London: Penguin, 1984.
Freud, Sigmund. *The Economic Problem of Masochism, Part XI (Addenda A, B, and C)*, translated by Joan Riviere. London: Hogarth Press, 1924.
Freud, Sigmund. *New Introductory Lectures on Psycho-Analysis and Other Works*, translated under the general editorship of James Strachey; in collaboration with Anna Freud; assisted by Alix Strachey and Alan Tyson. London: Vintage, Hogarth Press and the Institute of Psychoanalysis, 2001.
Fuller, Matthew, ed. *Software Studies: A Lexicon.* Cambridge, MA and London: MIT Press, 2008.
Geldard, Richard G. *Remembering Heraclitus.* New York: Lindisfarne, 2000.
Giddens, Anthony. *The Consequences of Modernity.* Stanford, CA: Stanford University Press, 1990.
Giesecke, Frederick E. *Hot-Water Heating and Radiant Heating and Radiant Cooling.* Austin, TX: Technical Books, 1947.
Gilbert, K. R. M. A. *Descriptive Catalogue of the Collection Illustrating Fire Fighting Appliances.* London: Science Museum, 1969.
Gotch, John Alfred. *The Growth of the English House: A Short History of its Architectural Development from 1100 to 1800.* London: Batsford, 1909.
Goudsblom, Johan. *Fire and Civilization.* London: Allen Lane, 1992.
Green-Hughes, E. *A History of Firefighting.* Ashbourne: Moorland, 1979.
Grosz, Elizabeth. *Architecture from the Outside: Essays on Virtual and Real Space.* Cambridge, MA: MIT Press, 2002.
Guo, Qinghua. *Chinese Architecture and Planning: Ideas, Methods, Techniques.* Stuttgart and London: Edition Axel Menges, 2005.
Halbwachs, Maurice. *On Collective Memory*, edited and translated by Lewis A. Coser. Chicago: University of Chicago Press, 1992.

Halbwachs, Maurice. *The Collective Memory*, translated by Francis J. Ditter, Jr and Vida Yazdi Ditter. New York: Harper & Row, 1980.
Haraway, Donna Jeanne. *Simians, Cyborgs, and Women: The Reinvention of Nature*. New York: Routledge, 1991.
Harrison, Rodney, ed. *Understanding the Politics of Heritage*. Manchester: Manchester University Press in association with the Open University, 2010.
Harrison, Rodney. *Heritage: Critical Approaches*. London: Routledge, 2013.
Hatherley, Owen. *Militant Modernism*. Winchester and Washington, DC: O Books, 2008.
Heidegger, Martin. *Being and Time*, translated by John Macquarrie and Edward Robinson. Malden, MA and Oxford: Blackwell, 2005.
Heidegger, Martin. *Contributions to Philosophy*, translated by Parvis Emad and Kenneth Maly. Bloomington and Indianapolis: Indiana University Press, 1999.
Heidegger, Martin. *The Origin of the Work of Art in Poetry, Language, Thought*, translated by Albert Hofstadter. New York: Harper & Row, 1971.
Heise, Ursula K., Jon Christensen and Michelle Niemann, eds, *The Routledge Companion to the Environmental Humanities*. London and New York: Routledge, 2017.
Hendrix Shannon, John, and Lorens Eyan Holm, eds, *Architecture and the Unconscious*. Farnham and Burlington, VT: Ashgate, 2016.
Hendrix Shannon, John. *Architecture and Psychoanalysis: Peter Eisenman and Jacques Lacan*. New York: Peter Lang, 2006.
Heschong, Lisa. *Thermal Delight in Architecture*. Cambridge, MA and London: MIT Press, 1979.
Hibberd, Shirley. *The Amateur's Greenhouse and Conservatory: A Handy Guide to the Construction and Management of Planthouses*. London, 1873.
Hill, Jennie, ed. *The Future of Archives and Recordkeeping: A Reader*. London: Facet, 2011.
Himmelblau, Coop. *Architecture is Now: Projects, (Un)buildings, Actions, Statements, Sketches, Commentaries 1968–1983*. London: Thames & Hudson, 1984.
Hollier, Denis. *Against Architecture: The Writings of Georges Bataille*, translated by Betsy Wing. Cambridge, MA and London: MIT Press, 1992.
Holm, Ivar. *Ideas and Beliefs in Architecture and Industrial Design: How Attitudes, Orientations and Underlying Assumptions Shape the Built Environment*. Oslo: Oslo School of Architecture and Design, 2006.
Hornsey, Richard. *The Spiv and the Architect: Unruly Life in Postwar London*. Minneapolis: University of Minnesota Press, 2010.
Ingram, Arthur. *A History of Fire-Fighting and Equipment*. London: New English Library, 1978.
Jenkinson, Hilary. *A Manual of Archive Administration*. London: Percy Lund, Humphries & Co., 1965.
Johnson, Walter Ralph. *Lucretius and the Modern World*. London: Gerald Duckworth & Co., 2000.
Jokilehto, Jukka. *A History of Architectural Conservation*. Oxford and Boston: Butterworth-Heinemann, 1999.
Kennett, Frances. *The Greatest Disasters of the 20th Century*. London: Marshall Cavendish, 1975.
Kerenyi, Karl. *Mnemosyne-Lesmosyne: On the Springs of 'Memory' and 'Forgetting'*. London: Routledge, 1977.
Koerner, Stephanie, and Ian Russell, eds, *Unquiet Pasts: Risk Society, Lived Cultural Heritage, Re-Designing Reflexivity*. Farnham and Burlington, VT: Ashgate, 2010.
Krell, Alan. *Burning Issues: Fire and Art in the Social Imagination*. London: Reaktion, 2011.
Kuberski, Philip. *The Persistence of Memory: Organism, Myth, Text*. Berkeley, CA and Oxford: University of California Press, 1992.
Lahiji, Nadir, and Daniel S. Friedman, eds, *Plumbing: Sounding Modern Architecture*. New York: Princeton Architectural Press, 1997.
Land, Nick. *The Thirst for Annihilation: Georges Bataille and Virulent Nihilism*. London and New York: Routledge, 1992.
Lash, Scott, and John Urry. *Economies of Signs and Space*. London: Sage, 1994.
Latour, Bruno. *Science in Action: How to Follow Scientists and Engineers Through Society*. Cambridge, MA: Harvard University Press, 1987.
Lawson, J. Randall. *NIST Technical Note 1628: A History of Fire Testing*. Gaithersburg, MD: National Institute of Standards and Technology, 2009.
Leach, Neil, ed., *Rethinking Architecture: A Reader in Cultural Theory*. New York: Routledge, 1997.
Leupen, Bernard. *Frame and the Generic Space: A Study into the Changeable Dwelling Proceeding from the Permanent*. Rotterdam: 2006.

Lincoln, Bruce. *Theorizing Myth: Narrative, Ideology, and Scholarship*. Chicago and London: University of Chicago Press, 1999.

Lissitzky-Küppers, Sophie. *El Lissitzky: Life, Letters, Texts*, translated by Helene Aldwinckle and Mary Whittall. London: Thames & Hudson, 1968.

Locke, John. *An Essay Concerning Human Understanding*. Oxford and New York: Oxford University Press, 2008 [1689].

Loudon, J. C. *Remarks on the Construction of Hothouses*. London, 1817.

Low, A. M. *Romance of Fire*. Worcester and London: Trinity Press, 1941.

Lucretius. *On the Nature of Things*, translated by Martin Ferguson Smith. Indianapolis and Cambridge: Hackett, 2001.

Luria, Aleksandr Romanovich. *The Mind of a Mnemonist: A Little Book About a Vast Memory*, translated by Lynn Solotaroff. New York: Basic Books, 1968.

Lyons, John W. *Fire*. New York: Scientific American Books, 1985.

Macarthur, John. *The Picturesque: Architecture, Disgust and other Irregularities*. London: Routledge, 2007.

Macdonald, Susan, Kyle C. Normandin and Bob Kindred. *Conservation of Modern Architecture*. Shaftesbury: Donhead, 2007.

Maly, Kenneth, and Parvis Emad. *Heidegger on Heraclitus: A New Reading*. New York: Edwin Mellen Press, 1986.

Marcus, Thomas A. *Buildings and Power: Freedom and Control in the Origin of Modern Building Types*. London and New York: Routledge, 1993.

Margalef, Ramon. *Perspectives in Ecological Theory*. Chicago and London: University of Chicago Press, 1968.

Mayer-Schoenberger, Viktor. *Delete: The Virtue of Forgetting in the Digital Age*. Princeton and Oxford: Princeton University Press, 2009.

Merewether, Charles, ed. *The Archive: Documents of Contemporary Art*. London: Whitechapel Gallery and Cambridge, MA: MIT Press, 2006.

Mitchell, James K., ed. *Crucibles of Hazard: Mega-Cities and Disasters in Transition*. New York: United Nations University, 1999.

Moneo, José Rafael. *Theoretical Anxiety and Design Strategies in the Work of Eight Contemporary Architects*, translated by Gina Cariño. Cambridge, MA and London: MIT Press, 2004.

Mullarkey, John, ed. *The New Bergson*. Manchester and New York: Manchester University Press, 1999.

Mullarkey, John. *Bergson and Philosophy*. Edinburgh: Edinburgh University Press, 1999.

Muñoz-Viñas, Salvador. *Contemporary Theory of Conservation*. Oxford: Butterworth-Heinemann, 2005.

Nadal, John B. *London's Fire Stations*. Huddersfield: Jeremy Mills, 2006.

Nesbit, Kate, ed. *Theorizing a New Agenda for Architecture: An Anthology of Architectural Theory 1965–1995*. New York: Princeton Architectural Press, 1996.

Nora, Pierre. *Realms of Memory: Rethinking the French Past*, translated by Arthur Goldhammer. New York: Columbia University Press, 1996.

Okwui, Enwezor, ed. *Archive Fever: Uses of the Document in Contemporary Art*. New York: International Center of Photography/Göttingen: Steidl, 2008.

Olson, Alan M., ed. *Myth, Symbol, and Reality*. Notre Dame, IN: University of Notre Dame Press, 1980.

Orbasli, Aylin. *Architectural Conservation: Principles and Practice*. Oxford: Blackwell, 2008.

Parikka, Jussi. *Digital Memory and the Archive*. Minneapolis: University of Minnesota Press, 2013.

Pfeifer, Günter and Brauneck, Per. *Courtyard Houses: A Housing Typology*, vol. 1, translated by Usch Engelmann. Berlin: Birkhäuser Verlag, 2008.

Pinto, John A. *Speaking Ruins: Piranesi, Architects, and Antiquity in Eighteenth-Century Rome*. Ann Arbor, MI: University of Michigan Press, 2012.

Plainiotis, Stelios, and Ji Yan, eds, *Design for Sustainability*. Beijing: China Architecture and Building Press, 2006.

Price, Nicholas Stanley, Mansfield Kirby Talley and Alessandra Melucco Vaccaro, eds, *Historical and Philosophical Issues in the Conservation of Cultural Heritage*. Los Angeles: J. Paul Getty Trust, 1996.

Pyne, Stephen. *Fire: A Brief History*. London: British Museum Press, 2001.

Quintiere, James G. *Fundamentals of Fire Phenomena*. Chichester: John Wiley & Sons, 2006.

Radstone, Susannah, and Bill Schwarz, eds, *Memory: Histories, Theories, Debates*. New York: Fordham University Press, 2010.

Rendell, Jane. *The Architecture of Psychoanalysis: Spaces of Transition*. London and New York: I. B. Tauris, 2017.
Richardson, Michael. *Georges Bataille*. London and New York: Routledge, 1994.
Ricoeur, Paul. *Memory, History, Forgetting*. Chicago: University of Chicago Press, 2004.
Ricoeur, Paul. *Time and Narrative*, vol. 3, translated by Kathleen Blamey and David Pellauer. Chicago: University of Chicago Press, 1988.
Rodovanovic, Milan, and Joao Fernando Pereira Gomes. *Solar Activity and Forest Fires*. New York: Nova Science Publishers, 2009.
Rossi, Aldo. *The Architecture of the City*, translated by Diane Ghirardo and Joan Ockman. Chicago: Graham Foundation for Advanced Studies in Fine Arts/New York: Institute for Architecture and Urban Studies, 1982.
Ruskin, John. *The Seven Lamps of Architecture*. Boston: Adamant Media, 2005.
Rykwert, Joseph. *On Adam's House in Paradise: The Idea of the Primitive Hut in Architectural History*. New York: Museum of Modern Art, 1972.
Rykwert, Joseph. *The Dancing Column: On Order in Architecture*. Cambridge, MA and London: MIT Press, 1996.
Rykwert, Joseph. *The Idea of a Town: The Anthropology of Urban Form in Rome, Italy and the Ancient World*. Princeton: Princeton University Press, 1976.
Samuel, Raphael. *Past and Present in Contemporary Culture*. London: Verso, 1999.
Samuel, Raphael. *Theatres of Memory*. London: Verso, 1994.
Schinaia, Cosimo. *Psychoanalysis and Architecture: The Inside and the Outside*, translated by Giuseppe Lo Dico. London: Karnac, 2016.
Scott, Charles E. *The Time of Memory*. New York: State University of New York Press, 1999.
Semper, Gottfried. *The Four Elements of Architecture and Other Writings*, translated by Harry Francis Mallgrave and Wolfgang Herrmann. Cambridge: Cambridge University Press, 1989.
Sennett, Richard. *The Craftsman*. London: Penguin, 2008.
Shacklock, Vincent, ed. *Architectural Conservation: Issues and Developments*, Shaftesbury: Donhead, 2006.
Shonfield, Katherine. *Walls Have Feelings: Architecture, Film and the City*. London and New York: Routledge, 2000.
Smith, D. N., ed. *Fire Safety Engineering*. Bedford: Fluid Engineering Centre, 1989.
Smith, D. N, ed. *Fire Safety Engineering: Proceedings of the 2nd International Conference on Fire Engineering and Loss Prevention in Offshore, Petrochemicals and Other Hazardous Applications*. Bedford: Fluid Engineering Centre, 1989.
Smith, Robert Rowland. *Death-Drive: Freudian Hauntings in Literature and Art*. Edinburgh: Edinburgh University Press, 2010.
Spieker, Sven. *The Big Archive: Art from Bureaucracy*. Cambridge, MA and London: MIT Press, 2008.
Spivey, Nigel Jonathan. *The Ancient Olympics*. Oxford: Oxford University Press, 2004.
Steadman, Philip. *The Evolution of Designs: Biological Analogy in Architecture and the Applied Arts*. Cambridge and New York: Cambridge University Press, 1979.
Stollard, Paul, and John Abrahams, *Fire From First Principles: A Design Guide to Building Fire Safety*. London and New York: E & F. N. Spon, 1991.
Stubbs, John H., and Emily Gunzburger Makas. *Architectural Conservation in Europe and the Americas: National Experiences and Practice*. Hoboken, NJ: John Wiley & Sons, 2011.
Stubbs, John H. *Time Honored: A Global View of Architectural Conservation: Parameters, Theory and Evolution of an Ethos*. Hoboken, NJ: John Wiley & Sons, 2009.
Surveyor-General of Prisons. *Report of the Surveyor-General of Prisons on the Construction, Ventilation, and Details of Pentonville Prison*. London, 1844.
Sylvester, Charles. *The Philosophy of Domestic Economy, As Exemplified in the Mode of Warming, Ventilating, Washing, Drying, and Cooking … Adopted in the Derbyshire General Infirmary, etc.* Nottingham, 1819.
Symposium on Fire Test Methods, Presented at the Fourth Pacific Area National Meeting. Los Angeles: American Society for Testing and Materials, 1963.
Treib, Marc, ed. *Spatial Recall: Memory in Architecture and Landscape*. New York: Routledge, 2009.
Tricker, Ray, and Rozz Algar. *Building Regulations in Brief*, fifth edition. Oxford: Butterworth-Heinemann, 2007.
Tung, Anthony M. *Preserving the World's Great Cities: The Destruction and Renewal of the Historic Metropolis*, New York: Clarkson-Potter, 2001.

Vanderveen, Bart H., ed. *Olyslager Auto Library: Fire-Fighting Vehicles 1840–1950*. London and New York: Frederick Warne, 1976.
Vidler, Anthony. *The Architectural Uncanny: Essays in the Modern Unhomely*. Cambridge, MA and London: MIT Press, 1992.
Virilio, Paul. *A Landscape of Events*, translated by Julie Rose. Cambridge, MA and London: MIT Press, 2000.
Vitruvius. *Vitruvius On Architecture*, translated by Thomas Gordon Smith. New York: Monacelli Press, 2003.
Jackson, W. E. *London's Fire Brigades*. London: Longmans, Green and Co., 1966.
Wallington, Neil. *Fire-Fighting: A Pictorial History*, Bristol: Parragon, 1997.
Wallington, Neil. *Images of Fire*. London: David and Charles, 1989.
Waterton, Emma, and Steve Watson, eds, *The Palgrave Handbook of Contemporary Heritage Research*. London: Palgrave Macmillan, 2015.
Watson and Pritchett Architects. *Plans, Elevations, Sections, and Description of the Pauper Lunatic Asylum … at Wakefield*. York, 1819.
Weatherill, Rob, ed. *The Death Drive: New Life for a Dead Subject?* London: Rebus Press, 1999.
Weinberger, Lois. *Debris Field: Explorations into the Decrepit 2010–2016*. Kassel: Documenta and Museum Fridericianum, 2017.
Weinrich, Harald. *Lethe, The Art and Critique of Forgetting*, translated by Steven Rendall. Ithaca, NY: Cornell University Press, 2014.
West, David Alexander. *The Imagery and Poetry of Lucretius*. Edinburgh: Edinburgh University Press, 1969.
Whitehead, Anne. *Memory*. London: Routledge, 2009.
Whittick, K. G. *The Party Wall: A Short History*. Godalming: Faculty of Party Wall Surveyors, 2007.
Wilton-Ely, John. *Piranesi as Architect and Designer*. New York and New Haven: Pierpont Morgan Library/Yale University Press, 1993.
Winnubst, Shannon, ed. *Reading Bataille Now*. Bloomington, IN: Indiana University Press, 2007.
Wright, Lawrence. *Home Fires Burning: The History of Domestic Heating and Cooking*. London: Routledge and Kegan Paul, 1964.
Xenophon. *Memorabilia*, translated by Amy L. Bonnett. Ithaca, NY and London: Cornell University Press, 2001.
Yates, Frances A. *The Art of Memory*. London: Pimlico, 2003.
Yatt, Barry D. *Cracking the Codes: An Architect's Guide to Building Regulations*. Chichester: John Wiley & Sons, 1998.
Zugmann, Gerald. *Blue Universe: Modelle zu Bildern machen. Transforming Models into Pictures. Architectural Models by Coop Himmelb(l)au*. Ostfildern: Hatje Cantz, 2002.
Zukin, Sharon. *Loft Living: Culture and Capital in Urban Change*. Baltimore and London: Johns Hopkins University Press, 1982.
Zurier, Rebecca. *The American Firehouse: An Architectural and Social History*. New York: Abbeville Press, 1982.

Journals, newspaper articles and magazines

Arden, Andrew. 'Editorial: Grenfell Tower: The Greatest Failure'. *Journal of Housing Law* 20: 5 (2017), 97–102.
Assmann, Jan. 'Collective Memory and Cultural Identity', translated by J. Czaplicka. *New German Critique*, 65 (1995), 125–34.
Bean, R., B. W. Olesen and K. W. Kim. 'History of Radiant Heating and Cooling Systems Part 1'. *ASHRAE Journal*, 52: 1 (2010), 40–7.
Bean, R., B. W. Olesen and K. W. Kim. 'History of Radiant Heating and Cooling Systems Part 2'. *ASHRAE Journal*, 52: 2 (2010), 50–55.
Bollas, Christopher. 'Architecture and the Unconscious'. *International Forum of Psychoanalysis*, 9: 1–2 (2000), 28–42.
Brooks, Graham. 'The Burra Charter: Australia's Methodology for Conserving Cultural Heritage'. *Places*, 8: 1 (1992), 84–8.

Bruegmann, Robert. 'Central Heating and Forced Ventilation: Origins and Effects on Architectural Design'. *Journal of the Society of Architectural Historians*, 37: 3 (October 1978), 143–60.
Burton, James. 'Bergson's Non-Archival Theory of Memory'. *Memory Studies*, 1: 3 (2008), 321–39.
Campbell, Aisling. 'Psychoanalysis and Architecture'. *Psychoanalysis and Space*, 4: 1 (Fall 2008), 1–13.
Colquhoun, Alan. 'Typology and Design Method'. *Perspecta*, 12 (1969), 71–4.
Connerton, Paul. 'Seven Types of Forgetting'. *Memory Studies*, 1: 3 (2008), 59–71.
Corris, Michael. 'White Out'. *Art Monthly*, 260 (October 2002).
Crang, Mike, and Penny S. Travlou. 'The City and Topologies of Memory'. *Environment and Planning D: Society and Space*, 19 (2001), 161–77.
Debate, 'Preservation, Contemporary Art, and Architecture'. *Future Anterior*, 4: 2 (Winter 2007), 71–85.
Featherstone, Mike. 'Archive'. *Theory Culture Society*, 23 (2006), 591–6.
Featherstone, Mike. 'Archiving Cultures'. *British Journal of Sociology*, 51: 1 (January/March 2000), 161–84.
Featherstone, Mike. 'The Flâneur, the City and Virtual Public Life'. *Urban Studies*, 35: 5–6 (1998), 909–25.
Fong, Benjamin Y. 'Death Drive Sublimation: A Psychoanalytic Perspective on Technological Development'. *Psychoanalysis, Culture and Society*, 18: 4 (2013), 352–67.
Fretin, Dominique, and Gilda Collet Bruna. 'Solar Energy: Harvesting the Sun. An Architectural Approach Through Le Corbusier's Groundings'. *Engevista*, 13: 1 (May 2011), 4–12.
Freud, Sigmund. 'Constructions in Analysis'. *The International Journal of Psycho-Analysis*, 19 (January 1938), 377–87.
Gallo, Emmanuelle. 'Jean Simon Bonnemain (1743–1830) and the Origins of Hot Water Central Heating'. *Proceedings of the Second International Congress on Construction History*, 1 (June 2006), 1043–60.
Gandy, Matthew. 'Cyborg Urbanization: Complexity and Monstrosity in the Contemporary City'. *International Journal of Urban and Regional Research*, 29: 1 (March 2005), 26–49.
Guggenheim, Michael. 'Building Memory: Architecture, Networks and Users'. *Memory Studies*, 2: 1 (2008), 39–53.
Harrison, Rodney. 'Beyond "Natural" and "Cultural" Heritage: Toward an Ontological Politics of Heritage in the Age of the Anthropocene'. *Heritage & Society*, 8: 1 (2015), 24–42.
Herscher, Andrew. 'Counter-Heritage and Violence'. *Future Anterior*, 3: 2 (Winter 2006), 25–33.
Holm, Lorens. 'Vignettes of Death: Architecture and the Death Drive'. *Critical Quarterly*, 49: 3 (2007), 31–59.
Jokilehto, Jukka. 'Preservation Theory Unfolding'. *Future Anterior*, 3: 1 (Summer 2006), 1–9.
Joyce, Patrick. 'The Politics of the Liberal Archive'. *History of the Human Sciences*, 12: 2 (1999), 35–49.
Kaplan, Edward K. 'Gaston Bachelard's Philosophy of Imagination: An Introduction'. *Philosophy and Phenomenological Research*, 33: 1 (September 1972), 1–24.
Kleinmann, Kent. 'Archiving/Architecture'. *Archival Science*, 1 (2001), 321–32.
Kontopodis, Michalis. 'Documents' Memories: Enacting Pasts and Futures at the School for Individual Learning-in-Practice'. *Memory Studies*, 2: 1 (2009), 11–26.
Kontopodis, Michalis. 'Editorial: Time. Matter. Multiplicity'. *Memory Studies*, 2: 1 (2009), 5–10.
Koolhaas, Rem. 'Cronocaos'. *Log*, 21 (Winter 2011), 119–23.
MacCannell, Juliet Flower. 'FreudSpace: Architecture in Psychoanalysis'. *Annual of Psychoanalysis*, 33 (2005), 93–108.
Mackenzie, Adrian. 'The Mortality of the Virtual: Real-Time, Archive and Dead-Time in Information Networks'. *Convergence*, 3: 2 (June 1997), 59–71.
Manoff, Marlene. 'Theories of the Archive from Across the Disciplines'. *Libraries and the Academy*, 4: 1 (2004), 9–25.
Martín Hernández, Manuel. 'Architecture from Architecture: Encounters Between Conservation and Restoration'. *Future Anterior*, 4: 2 (Winter 2007), 62–9.
Moneo, Rafael, 'On Typology'. *Oppositions*, 13 (Summer 1978), 21–45.
Nora, Pierre. 'Between Memory and History: Les Lieux de Mémoire' translated by Marc Roudebush. *Representations*, 26 (Spring 1989), 7–24
O'Gorman, Marcel. 'Bernard Stiegler's Pharmacy: A Conversation'. *Configurations*, 18: 3 (2010), 459–76.
Olick, Jeffrey K. 'Collective Memory: A Memoir and Prospect'. *Memory Studies*, 1: 1 (2008), 23–9.
Otero-Pailos, Jorge. 'Editorial: Chance Architecture'. *Future Anterior*, 3: 2 (Winter 2006), ii–viii.

Otero-Pailos, Jorge. 'Editorial: Creative Agents'. *Future Anterior*, 3: 1 (Summer 2006), ii–vii.
Otero-Pailos, Jorge. 'Editorial: Echoing: How to Situate the New'. *Future Anterior*, 2: 1 (Summer 2005), ii–vii.
Otero-Pailos, Jorge. 'Editorial: Historic Provocation: Thinking Past Architecture and Preservation'. *Future Anterior*, 2: 2 (Winter 2005), ii–vi.
Perraudin, Jean François, 'A Non-Bergsonian Bachelard'. *Continental Philosophy Review*, 41: 4 (2008), 463–79.
Picart, Caroline Joan ('Kay') S. 'Metaphysics in Gaston Bachelard's "Reverie"'. *Human Studies*, 20: 1 (January 1997), 59–73.
Picon, Antoine. 'Anxious Landscapes: From the Ruin to Rust', translated by Karen Bates. *Grey Room*, 1 (Fall 2000), 64–83.
Poole, Ross. 'Memory, History and the Claims of the Past'. *Memory Studies*, 1:1 (2008), 149–66.
Power, Nina. 'Bachelard *Contra* Bergson, Ancient Atomism and the Debate Over Continuity'. *Angelaki*, 2: 3 (December 2006), 117–22.
Rapaport, Herman. 'Archive Trauma'. *Diacritics*, 28: 4 (Winter 1998), 68–81.
Raz, Yosef. 'Traces of War: Memory, Trauma, and the Archive in Joseph Cedar's *Beaufort*'. *Cinema Journal*, 50: 2 (Winter 2011), 61–83.
Reisner, Gavriel. 'Revisioning the Death-Drive: The Compulsion to Repeat as a Death-in-Life'. *Psychoanalytic Review*, 101: 1 (February 2014), 39–69.
Rendell, Jane. 'Working Between and Across: Some Psychic Dimensions of Architecture's Inter- and Transdisciplinarity'. *Architecture and Culture*, 1: 1 (2013), 128–40.
Riegl, Alois. 'The Modern Cult of Monuments: Its Character and its Origin'. *Oppositions*, 25 (Fall 1982), 21–56.
Roediger, Henry L., and James V. Wertsch. 'Creating a New Discipline of Memory Studies'. *Memory Studies*, 1: 1 (2008), 9–22.
Russell, Conrad. 'Fictive Time: Bachelard on Memory, Duration and Consciousness'. *Kronoscope*, 5: 1 (2005), 3–18.
Savoy, Eric. 'Aspern's Archive'. *Henry James Review*, 31: 1 (Winter 2010), 61–7.
Schmetterling, Astrid. 'Archival Obsessions: Arnold Dreyblatt's Memory Work'. *Art Journal*, 66: 4 (Winter 2007), 70–83.
Schwartz, Joan M., and Terry Cook. 'Archives, Records, and Power: The Making of Modern Memory'. *Archival Science*, 2 (2002), 1–19.
Sperber, Esther. 'Sublimation: Building or Dwelling? Leowald, Freud, and Architecture'. *Psychoanalytic Psychology*, 31: 4 (2014), 507–24.
The Times, 19 November 1987, 'King's Cross Fire, 18 November 1987'.
Till, Karen E. 'Artistic and Activist Memory-Work: Approaching Place-Based Practice'. *Memory Studies*, 1: 1 (2008), 99–113.
Transactions of the Society, Instituted at London, for the Encouragement of Arts, Manufactures, and Commerce, 24 (1806).
Webb, Sam. 'Councils Should Learn from King's Cross Fire'. *Architects Journal*, 230 (May 2009).
Wigley, Mark. 'Unleashing the Archive'. *Future Anterior*, 2: 2 (Winter 2005), 11–15.
Williams, Jessica. 'Introduction: Theories Toward a Critical Practice'. *Future Anterior*, 2: 2 (Winter 2005), vii–xii.
Wills, David. 'Post/Card/Match/Book/Envois/Derrida'. *Substance*, 43 (1984), 19–38.
Wisniak, Jaime. 'Matches: The Manufacture of Fire'. *Indian Journal of Chemical Technology* 12 (May 2005), 369–80.
Wright, Patrick, and Jeremy Davies. 'Just Start Digging: Memory and the Framing of Heritage'. *Memory Studies*, 3: 3 (2010), 196–203.

Reports

'The Charter of Krakow 2000: Principles for Conservation and Restoration of Built Heritage', 2000. Accessed 6 December 2017. http://smartheritage.com/wp-content/uploads/2015/03/KRAKOV-CHARTER-2000.pdf.
English Heritage, 'Conservation Principles, Policies and Guidance', 2008. Accessed 15 August 2018. https://www.historicengland.org.uk/images-books/publications/conservation-principles-sustainable-management-historic-environment.

English Heritage, 'PPS5: Planning for the Historic Environment: Historic Environment Planning Practice Guide', 2010. Accessed 12 October 2017. http://www.communities.gov.uk/publications/planningandbuilding/pps5.

English Heritage, 'Research Strategy', December 2016. Accessed 15 August 2018. https://content.historicengland.org.uk/images-books/publications/research-strategy/research-strategy.pdf/.

English Heritage, 'Understand Historic Buildings: A Guide to Good Recording Practice', 2006. Accessed 22 February 2019. https://historicengland.org.uk/images-books/publications/understanding-historic-buildings/heag099-understanding-historic-buildings/.

HM Government, *The Building Regulations 2010*, Fire Safety, Approved Document B, Volume 1: Dwellinghouses (2006 edition incorporating the 2010 and 2013 amendments) (London: NBS, 2011).

Home Office, 'Response Times to Fires Attended by Fire and Rescue Services: England, April 2016 to March 2017', 25 January 2018. Accessed 20 June 2018. https://www.gov.uk/government/statistics/response-times-to-fires-attended-by-fire-and-rescue-services-england-april-2016-to-march-2017.

UNESCO, 'The Nara Document on Authenticity', 1994. Accessed 2 August 2018. https://whc.unesco.org/document/116018.

Websites

BBC News. 'Grenfell Tower: What Happened'. Accessed 18 June 2018. https://www.bbc.co.uk/news/uk-40301289.

Bonde, Lisbeth. 'Video Pyromaniacs and Pictures of Fire Related Works'. Accessed 26 August 2018. https://piodiaz.wordpress.com/2007/02/20/video-pyromaniacs/.

Close-Up. '1 February 2011: Abandoned Archives and Forgotten Histories Remembered'. Accessed 23 August 2018. https://www.closeupfilmcentre.com/film_programmes/2011/abandoned-archives-and-forgotten-histories-remembered/.

Evening Standard. David Cohen, 'Leaked Grenfell Dossier Reveals how Disastrous Refurbishment Turned Tower into a "Tinderbox"', 16 April 2018. Accessed 6 August 2018. https://www.standard.co.uk/news/london/shock-grenfell-dossier-reveals-disastrous-refurbishment-turned-tower-into-a-tinderbox-a3814866.html.

Historic England. 'Cultural Landscapes'. Accessed 23 August 2018. https://historicengland.org.uk/listing/what-is-designation/local/conservation-areas/.

Historic England. 'Definition: Heritage Asset'. Accessed 21 August 2018. https://historicengland.org.uk/advice/hpg/hpr-definitions/h/536274/.

Historic England. 'Identification and Designation of Heritage Assets'. Accessed 21 August 2018. https://historicengland.org.uk/advice/hpg/has/.

Historic England. 'Listed Buildings'. Accessed 23 August 2018. https://historicengland.org.uk/listing/what-is-designation/listed-buildings/.

Historic England. 'We', 2018. Accessed 23 August 2018. https://historicengland.org.uk/.

London Fire Brigade. 'Firefighters of the Second World War'. Accessed 26 September 2018. https://www.london-fire.gov.uk/museum/history-and-stories/firefighters-of-the-second-world-war/.

Philippe Rahm Architectes. 'Digestible Gulf Stream'. Accessed 1 March 2018. http://www.philipperahm.com/data/projects/digestiblegulfstream/index.html.

Pierre Nora. 'Reasons for the Current Upsurge in Memory', *Eurozine*. Accessed 19 April 2018. http://www.eurozine.com/articles/2002–04-19-nora-en.html.

Royal Voluntary Service. 'Our History'. Accessed 26 September 2018. https://www.royalvoluntaryservice.org.uk/about-us/our-history.

UNESCO. 'Dresden is Deleted from UNESCO's World Heritage List'. Accessed 8 February 2018. http://whc.unesco.org/en/news/522.

UNESCO. 'Historic Centre of Warsaw'. Accessed 30 April 2018. http://whc.unesco.org/en/list/30.

UNESCO. 'Oral Traditions and Expressions including Language as a Vehicle of the Intangible Cultural Heritage'. Accessed 4 July 2018. http://www.unesco.org/culture/ich/index.php?pg=00053.

Archive material

Barbara Lane (April 12, 2018), 'Grenfell Tower – fire safety investigation: The fire protection measures in place on the night of the fire, and conclusions as to: the extent to which they failed to control the spread of fire and smoke; the extent to which they contributed to the speed at which the fire spread', Grenfell Tower Inquiry hearing, 12 April 2018, https://assets.grenfell-towerinquiry.org.uk/documents/Dr%20Barbara%20Lane%20report%20-%20section%20 5%20-%207.pdf, accessed 9 February 2019.

Daniel Brown, Witness statement, Grenfell Tower Inquiry hearing, 4 June 2018, https://www.grenfelltowerinquiry.org.uk/hearings, accessed 31 July 2018.

London Fire Brigade, 'Grenfell Tower Inquiry: Opening Statement of the London Fire Brigade', 18 May 2018, https://www.grenfelltowerinquiry.org.uk/sites/default/files/documents/London%20Fire%20Brigade%20opening%20statement.pdf, accessed 3 August 2018.

London Fire Brigade Chief Officer Captain R. N. (Rtd.), Letter, 7 October 1905, on water supplies. London Metropolitan Archives LCC/FB/GEN/06.

Niamh Nic Daéid, Witness statement, Grenfell Tower Inquiry hearing, 19 June 2018, https://www.grenfelltowerinquiry.org.uk/evidence/professor-niamh-nic-daeids-presentation, accessed 9 February 2019.

Films

Toute la mémoire du monde, directed by Alain Resnais. France: Film de la Pléiade, 1957, 35mm.

Index

Page numbers in italics refer to figures

www.ingramcontent.com/pod-product-compliance
Lightning Source LLC
LaVergne TN
LVHW052017300826
846742LV00020B/84